ciples of statistics
ing economic dat~
n analysing
sting ec

£ 7·10

D1171255

About the Authors

Anne C. Mayes, (B.Sc. Wales) graduated from University College, Cardiff in Pure Mathematics and Economics. She moved to Exeter as a Lecturer in Economic and Social Statistics in 1968.

David G. Mayes, (M.A. Oxon, PhD. Bristol) having completed his PhD. on The Effects of Alternative Trade Groupings on the U.K., at Bristol, moved to Exeter as a Lecturer in Economic and Social Statistics in the Department of Economics in 1971. He has published articles on economic integrations, trade in agricultural products and estimation problems in international trade, and is currently engaged in research in these topics and in the acceptability of fertility regulating methods.

Introductory Economic Statistics

Introductory Economic Statistics

ANNE C. MAYES
and
DAVID G. MAYES

*Department of Economics,
University of Exeter*

JOHN WILEY & SONS
London · New York · Sydney · Toronto

Copyright © 1976 by John Wiley & Sons, Ltd

Library of Congress Cataloging in Publication Data:

Mayes, Anne C.
    Introductory economic statistics.

  Includes bibliographies.
    1. Statistics. I. Mayes, David G., joint author.
II. Title.
HA29.M297   519.5   75—15838
ISBN 0 471 58031 7 (Cloth)
ISBN 0 471 58111 9 (Pbk)

Typeset in IBM Century by Preface Limited, Salisbury, Wilts and printed in the United States of America.

Preface

The writing of this book has drawn on our experience in teaching and research in the subject and our conversations with friends and colleagues. We are, however, especially grateful to Desmond Corner who read the entire draft and commented in detail on the work. We would also like to thank the two reviewers appointed by the publisher for their suggestions for improvements and their watchful eye for accidental errors of substance. The first part of the book benefitted from the careful scrutiny and examination of John Black, but without the help of the publishers, in the face of typing difficulties and other delays our energies might well have flagged. We have not deliberately drawn on the work of others without acknowledgement, and we hope that any random similarities will be taken in a spirit of self-satisfaction. Finally we make the usual disclaimer for any remaining errors — our colleagues should have spotted them!

Those passages in the book which are marked by continuous stars provide additional or more detailed information and may be omitted if desired.

We acknowledge with many thanks the permission of the Controller of H.M.S.O. to use the large number of Official Statistics we have included.

A.C.M.
D.G.M.
1975

Contents

viii

1 Introduction

Although some statistics is taught to almost all students of economics, the importance of statistics in economics is often not brought out from the beginning. Statistics forms the backbone of the application of economic theory. It is used in the description of economic phenomena, the estimation of economic relationships, the testing of economic theories and the prediction and forecasting of economic variables. In this book we consider these four topics:

Description

Estimation

Testing

Prediction

In doing this we shall provide the basic knowledge that any economist will require if he is to read the articles and books which are used in the study of economics. Thus in some respects we go beyond what is normally considered a first-year text to provide adequate coverage for the student who does not wish to specialize in statistics or econometrics, although we hope that through the reading of this text he will be encouraged to take his studies further.

1.1 The Concept of an Economic Model
In suggesting that our study be confined to four topics we have omitted an overriding precondition, namely that economic theory be formulated in such a way that we have relationships to estimate and hypotheses to test. If this were not so statistics would be a mere descriptive tool. This essentially means that we must make use of a little elementary mathematics. Let us take the determination of the level of investment in the economy as an example. A simple model would be to suggest that investment runs at a constant level. We might

formulate this as

$$I = a \tag{1}$$

where I stands for investment and a is the constant level. A more sophisticated hypothesis might be to suggest that part of investment is determined by the level of interest rates. This we can write as

$$I = a + f(R) \tag{2}$$

where R stands for the level of interest rates and $f(\ )$ for 'is a function of'. By using this formulation we are not prejudging the particular form of the relationship. If it is our opinion that the relationship is linear then we can write

$$I = a + bR \tag{3}$$

In this case, as R changes, I changes by the proportion, b, of the change in R.

1.2 The Role of Statistics in Economic Models

Since equation (3) is a specific formulation of an economic relationship, we could now use statistics to estimate what the values of a and b are in practice. We consider in Chapter 6 how we might undertake this sort of exercise for the British economy over the last few years. Thus we can use statistics to provide actual values for the relations that economic theory suggests.

A further, and perhaps more interesting, step is to see if we think that the values obtained for a and b actually tell us anything useful. Let us say that the value of b which we calculate turns out to be a rather small number. This could mean one of two things; either that the rate of interest does not really have any effect on the level of investment, or that for any change in interest rates the change in investment is much smaller. Since investment and interest rates are measured in different units, £thousand million and % for example, a small value for b may not be implausible. We really want to know whether (1) or (3) is the case. In other words we wish to test the hypothesis that b is not really different from zero. Chapters 3 and 4 show the criteria we can use to decide whether to accept or reject an hypothesis about a model which we have estimated.

Having estimated a relationship and decided to accept it we can go on to suggest what might happen in other similar situations. We might for example suggest that if (3) has held over a period of ten years that it is likely that it also held in the following year. If we know what the value of R was in the eleventh year we can 'forecast' what I, investment, was at that time by multiplying R by b and adding a. Obviously, since we know what the actual value of I was in the eleventh year we can see how good our forecast is. In Chapters 6 and 7 we consider the problems of predicting and forecasting the values of economic variables from our estimated models.

We conclude the main part of the book in Chapter 8 by looking forward to the problems of putting the techniques described here into practice. A major difficulty underlying economic analysis is that we are always faced by several economic forces in any one situation. It is the task of the economist to disentangle these forces. This chapter brings together the ideas presented earlier in the book to show how the problems of the application of economic theory are tackled, so that the reader can be left with sufficient understanding to proceed with the study of economic literature.

However, we begin by considering a simpler problem. We can use statistics to describe what has happened to the variables during the last few years. The purpose of descriptive statistics is to summarize and present data in a way to make them more readily comprehensible. We shall begin by considering descriptive statistics in Chapter 2.

Finally, we have included as an Appendix a chapter on the use and construction of index numbers. Since these variables are widely used in economics we felt our text would not be complete without an explicit discussion of them. The chapter is self-contained so it can be read at any stage during the reading of the rest of the book or even entirely alone. The text can be read completely without reading the Appendix, but we recommend that it be read either after Chapter 2 or before starting Chapter 6.

2 Distributions and the Meaningful Description of Distributions

2.1 Groups of Events

There are a vast number of events which form the data on which economic analysis is based. Although every event is separate, economics is concerned with generalization about economic behaviour, and we can consider most events as members of some sort of group. For example, we can talk about 'earned incomes in the United Kingdom in 1973' as a group, whose members are individual incomes. We can talk about 'earned income'. Such a concept is called a *variable*. It is a variable because it can hold a variety of different values. We can contrast this with the logarithm of 100, which is a *constant*. No matter how many times we calculate log 100 it has the same value, whereas we can observe many different earned incomes.

In this chapter and throughout the book we are concerned with these two basic concepts of a group of events and a variable. We shall refer to these groups of events as *populations*. Statistics can be defined as the branch of scientific method which deals with the data obtained by counting or measuring the properties of populations of natural phenomena (see Kendall and Stuart, Vol. 1, p. 2). The subject 'Statistics' must be distinguished from the term 'a statistic', which is defined as a function of the observations in a sample drawn from some population. We shall deal with sampling in Chapter 3 and only consider descriptive statistics of the whole population in this chapter. The values of the variable, earned income, form the population of earned incomes.

If we take a small population we can deal with each of its members. Let us say that the variable we wish to use is the amount spent by members of our family yesterday. Assuming there are four of us we can record these amounts as follows: 2 p, 20 p, 75 p and £2.25. We could also record the values in the population of expenditure by the members of the family over the last week as there would be only twenty-eight values to write down. However, if we were to extend the time period to the past year the situation would become unmanageable.

We need to devise some easier method of setting out these events, so that we can grasp their main features. If we set out the values of the variable in order of magnitude, we can divide up the values into intervals.

Let us proceed diagrammatically.

The line AB includes all values of daily expenditure during the past year. In fact the smallest value actually recorded was zero and the largest £100. The range of values is thus £100. A simple division of this range into intervals would be into 10 equal intervals of £10 each: up to £10, greater than £10 but not more than £20 and so on. We can set up a table of these intervals as in Table 2.1, and show the distribution of the values over their range. The intervals are usually referred to as *class intervals* and the distribution as a *frequency distribution*. The frequencies are the number of observations in each class.

In dividing the range up into class intervals we must bear in mind that not all variables can hold every value within the range. Some variables are *discrete* in that they have only specific possible values. For example, the number of people in a household can only be a whole number. Age on the other hand is *continuous* in that it can hold any value, however small a unit we wish to use. Most economic variables, such as expenditure in this example, can be treated as continuous, although they are discrete at some level, such as a halfpenny.

Table 2.1
Frequency distribution of daily expenditure over the last year

Interval		Number of observations
greater than	not greater than	
	£10[a]	1,397
£10	£20	25
£20	£30	13
£30	£40	7
£40	£50	5
£50	£60	2
£60	£70	4
£70	£80	2
£80	£90	3
£90	£100	2
Total		1,460

[a]This class includes zero.

The frequency distribution of daily expenditure shows, as we might expect, that the vast majority of values fall in the range £0 to £10. In fact, because so many of the values fall in the first class interval, we have not learnt a great deal about the distribution of values. Obviously there are a number of points that we must bear in mind in choosing class intervals:

(i) Trivially, every member of the population must be included in a class interval, and no member may be included more than once. Thus class intervals must not overlap. If we were to define classes as £10—£20, £20—£30 and so on, then a value of £20 would fall into two classes. Secondly we can make sure that no value is excluded by having the first and last classes open-ended. For example, 'less than £6' for a first class includes all observations less than £6 down to minus infinity, and '£100 and above' as a last class interval will include all observations from £100 to plus infinity. (Clearly in this case the lowest class is 'closed' at 0 as expenditure cannot be negative, and the highest class will have some reasonable finite limit.)

(ii) As far as possible class intervals should be of equal range, so that it is easier to compare frequencies between classes. If equal class intervals would not show the way the population is distributed clearly, then large class intervals should be a simple multiple of the smallest interval. Thus, for example, if the smallest interval were £5, then we might choose £10 and £50 (multiples of 2 and 10 respectively) as larger intervals.

In Table 2.2 we have shown how such a frequency distribution is drawn up for the estimated net wealth of individuals in Great Britain in

Table 2.2
Estimated net wealth of individuals in Great Britain, 1960

Ranges of net wealth				
over £'000	not over £'000	Number of cases '000	Frequency per £1000 interval	Cumulative frequency '000
	1	8,671	8,671	8,671
1	3	5,582	2,791	14,253
3	5	2,361	1,181	16,614
5	10	1,647	329	18,261
10	15	481	96	18,742
15	20	221	44	18,963
20	25	126	25	19,089
25	50	250	10	19,339
50	100	99	2	19,438
100	200	32	0	19,470
200		13	?	19,483
Total		19,483		

Source: *Annual Abstract of Statistics*

1960. The smallest interval used is £1,000, and the remainder are multiples of 2, 5, 25, 50 and 100 times it. All the possible range is covered and there are no overlaps in the classes. As we might expect, the greatest frequency lies in the class intervals at the lower end of the range, a fact which is emphasized by the relatively smallness of these first intervals. We have thus overcome the problem of lack of information given by the very high frequency in the first interval in Table 2.1.

2.2 Diagrammatic Representation of Distributions

2.2.1 Histograms

Two diagrammatic methods are frequently used for showing the shape of frequency distributions. The first of these is the histogram. Both measures are forms of graphs of frequency against the values of the variable. The horizontal axis is the value of the variable, but in order to get a vertical axis which is consistent we must use frequency per fixed interval, otherwise we would get totally different shapes of histograms depending upon the intervals we used. In effect the histogram represents frequency by area. Using Table 2.2, we must first of all decide upon the size of our fixed interval over which we shall calculate the frequencies. The obvious choice is the smallest interval of £1,000, and approximate frequencies per £1,000 interval are shown in the third column of the table. The frequencies are approximate because we are assuming that the values are evenly distributed across the class interval. A quick look at Figure 2.1 shows that this assumption requires a

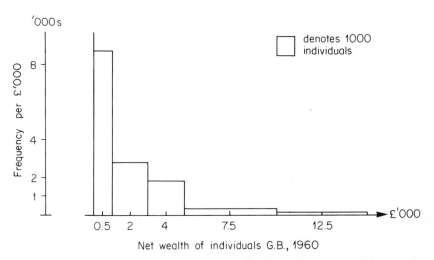

Figure 2.1 Histogram of net wealth of individuals in Great Britain, 1960. Note that the vertical axis is not normally included, but is given here to make it clear how the size of the histogram is calculated. Frequency is shown by area. Source: *Annual Abstract of Statistics*

considerable degree of approximation. It is apparent that above
£15,000 we cannot really distinguish the frequency/£1,000 from the
horizontal axis, and above £200,000 we cannot calculate the fre-
quency/£1,000 as we do not know the upper limit of the class. Under
some circumstances we can make an intelligent guess about the likely
upper limit, but there is usually no need to do so. If the lower limit of
the first class is unstated we may also have a similar problem there.

2.2.2 Frequency Polygons
The second method of graphical representation is a frequency polygon.
Again the vertical axis is frequency per class unit and the horizontal,
the variable values. In this case, however, we plot the points of
frequency over the midpoint of the class interval instead of drawing a
horizontal line across the entire interval, and the join the plotted points
with a straight line. (It is also conventional to continue the line to the
horizontal axis at each end of the distribution. This is drawn to a point
which would be the midpoint of a hypothetical class interval at either
end of the actual intervals each of the same width as the interval next
to it.) A frequency polygon of the distribution of estimated numbers of
female employees in Great Britain at mid-June, 1960 is drawn in
Figure 2.2 from the information in Table 2.3.

2.3 The Description of Distributions
We have drawn up Figure 2.2 without any comment on what it shows.
A simple description might run as follows. The maximum number of
employees lies in the 18- and 19-year-old category when almost all girls
will have left school, but many will not yet have married. Over the
previous three years, 15, 16 and 17, the frequency is next highest, the
lower level perhaps being accounted for by those still at school. After

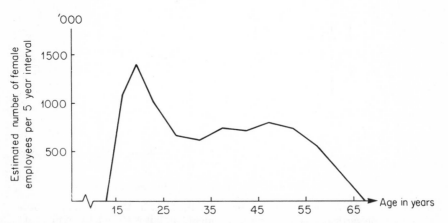

Figure 2.2   Frequency polygon of estimated numbers of female employees,
mid-June 1960, Great Britain. Source: *Annual Abstract of Statistics*

Table 2.3
Estimated numbers of female employees — mid-June,
1960 (G.B.)

Age years	Midpoint years	Frequency '000	Frequency per 5 years '000
Under 18	16.5	666	1,110[a]
18 and 19	19.0	528	1,320
20—24	22.5	1,047	1,047
25—29	27.5	678	678
30—34	32.5	627	627
35—39	37.5	750	750
40—44	42.5	731	731
45—49	47.5	814	814
50—54	52.5	745	745
55—59	57.5	568	568
60—64	62.5	278	278
65 and over	?	168	?

[a]15 years assumed as lowest age.

Source: *Annual Abstract of Statistics*

19 the numbers drop away to a minimum in the 30—34 age range as females marry and become mothers. After this it appears that some females tend to return to work and a slight increase is recorded up to a second maximum in the 45—49 age range, after which the numbers drop off rapidly with early retirement and (official) retirement at 60.

2.4 Measures of Central Tendency
2.4.1 The Mode
The main feature which we have picked out is the highest point of the distribution. Any point of maximum frequency is called a mode. Thus the distribution has three modes, one over the range 18 and 19 years, the second 35—39 years and the third 45—49 years. Since the distribution is divided into classes we have modal ranges rather than modes. In Figure 2.1 the modal range of the distribution of net wealth of individuals in Great Britain in 1960 is the first interval, £0—£1,000. There is no particular name for points or ranges of minimum frequency.

2.4.2 The Arithmetic Average
Another simple measure we can use is the arithmetic average. This average or mean is a measure of the centre of the distribution, around which all the other points lie. Its calculation is trivial if we know all the individual values of the population. If we call the variable X and the number of members of the population N then the arithmetic mean, $\mu = \Sigma X/N$.

★ For those not conversant with the symbol Σ, Σ means 'sum of', and

★ refers to the variable(s) following it. Strictly, ΣX is short for
★
★
★ $$\sum_{i=1}^{i=N} X_i = X_1 + X_2 + \ldots + X_N$$
★
★
★ where each of the X_is are the individual members of the population.
★ Thus ΣX is the sum of X_i from $i = 1$ to $i = N$. If any of the items i,
★ 1 or N are obvious then they are omitted; thus
★
★
★ $$\Sigma X, \ \Sigma_i X_i, \ \Sigma_i X_i, \ \sum_{i=1}^{N} X_i \quad \text{and} \quad \sum_{i=1}^{i=N} X_i$$
★
★ are all equivalent.

However, in the case of a frequency distribution we do not know the individual X_i. Now, $\mu = \Sigma fX/\Sigma f$, where f is the class frequency and X the midpoint of the class. But since we are assuming that values of the variable are evenly distributed across each class interval we are only estimating the mean of the population rather than calculating its exact value. If we return to Table 2.1 the mean of the distribution is £6.2 and the calculation is shown in Table 2.4.

2.4.3 Mathematical Expectation
We can take the use of the mean a little further in that it is called the expected value of the variable. Thus

$$E[X] = \mu$$

Table 2.4
Calculation of the mean and median

Daily Expenditure During the Past Year

Class interval		Midpoint (X)	Frequency (f)	(fX)	Cumulative frequency
greater than	not greater than				
	£10[a]	5	1,397	6,985	1,397
£10	£20	15	25	375	1,422
£20	£30	25	13	325	1,435
£30	£40	35	7	245	1,442
£40	£50	45	5	225	1,447
£50	£60	55	2	110	1,449
£60	£70	65	4	260	1,453
£70	£80	75	2	150	1,455
£80	£90	85	3	255	1,458
£90	£100	95	2	190	1,460
Total			1,460	9,120	

$\Sigma fX/\Sigma f = 9,120/1,460 = £6.20$

[a]Includes zero.

where μ is the mean of X and $E[\ ]$ refers to 'the expected value of'. (Strictly, in calculating $E[X]$ we should deal in terms of relative frequency; thus $E[X] = \Sigma pX$ where $p = f/\Sigma f$.)

2.4.4 The Median

The remaining measure of location in common use is the median. Like the mean, this is a measure of the centre of the distribution. When all the values of the distribution are set out in order of magnitude the median is the middle value. There are as many values greater than the median as there are less than it. If the total number of values in the population is even, then the median is the mean of the two middle observations. Thus if there are 21 observations, the median is the value of the 11th observation, but if there are 20 it is the average of the 10th and 11th observations.

If we are using grouped data then we can easily identify the median class by looking at the cumulative frequency distribution. This is shown for our expenditure example in the last column of Table 2.4, where the cumulative frequency is the sum of all frequencies from the first up to and including the class in question. Thus the cumulative frequency for the first class is merely itself, and that for the second the sum of the frequencies of the first two classes and so on. When we are dealing with grouped data, we can only estimate the median as we do not know the exact distribution of observations in the median class, hence we use $\Sigma f/2$ as our median 'observation' without any adjustment for an odd or even numbered total frequency. Therefore since there are 1,460 observations altogether the $1,460/2 = 730$th observation is the median one. The first class for which the cumulative frequency includes the median observation is clearly the class which contains the median. In this case it is obviously the first class, and we can see from the distribution of wealth in Table 2.2 that there the median lies in the second class, over £1,000 but less than £3,000. However, we can obtain a precise value for the median by making use of the assumption of a uniform distribution of observations within each class. If the median observation is sixth out of twenty members of a class then it is 6/20th of the way along the range of the class as well. Let us set this up formally defining

M as the median.
F as the cumulative number of observations in all classes below the median class,
f_M as the number of observations in the median class,
W as the width of the class,
L as the value of the lower bound of the class.

We can calculate the proportion of the way through the class as $[\frac{1}{2}(\Sigma f) - F]/f_M$ and then multiply by W to convert this into values and add it to L, the lower bound of the class. Thus

$$M = L + W \left(\frac{\frac{1}{2}(\Sigma f) - F}{f_M} \right)$$

In the expenditure example

$M = 0 + 10(730 - 0)/1,397 = £5.23$

and in the wealth example

$M = 1,000 + 2,000(9,742 - 8,671)5,582 = £1,384$
to the nearest £.

2.4.5 The Measures Compared

The mode, mean and median all provide information about the centre of a distribution, but this information is only the same in certain specific circumstances. Usually they tell us different things and one or other of them will be more appropriate to our particular purpose.

Let us say for example that a charity has the opportunity to use the parking area in front of a supermarket to collect money for any one hour they like to name on a Saturday morning. If the charity wishes to raise as much money as possible it will pick the hour in which the greatest number of people usually visit the supermarket. Thus if it looks at the distribution of customer arrivals over time it will want to determine the modal hour, not the time by which 50% of the customers have arrived nor still less the average time at which people shop. The average time may be severely distorted by a large number of people who shop on the way to work. We can see from Table 2.5 that the

Table 2.5
Distribution of customer arrivals on a Saturday morning

Hour[a]	Number of arrivals (f)	Midpoint (X)	(fX)	Cumulative number of arrivals
7 to 8	50	7.5	375	50
8 to 9	150	8.5	1,275	200
9 to 10	50	9.5	475	250
10 to 11	100	10.5	1,050	350
11 to 12	200	11.5	2,300	550
12 to 1	300	12.5	3,750	850
Total	850		9,225	

Modal class is 12 to 1 (ignoring the minor mode of 8 to 9)
Mean is $\Sigma fX/\Sigma f$ = 9,225/850 = 10.85 or 10.51 am (as 0.85 of an hour is 51 minutes)
Median arrival 850/2 = 425th which clearly occurs between 11 and 12
Median = $11 + (425 - 350)/200$ = 11.38 or 11.23 am.

[a]Hour includes lower but not upper limit.

modal hour is 12 to 1 (ignoring the minor modal range of 8 to 9), the median lies in the hour 11 to 12 and the mean lies in the hour 10 to 11.

We would want to look at the median in response to such questions as: 'Am I earning more than most people?'. The main table in *The New Earnings Survey* (published annually by H.M.S.O.) shows the distribution of earned incomes in each year, and calculates the median for just this sort of reason. If, for example, I had earned £40 per week in April, 1972 as a male non-manual worker I would have earned more than most people, median = £38.5, but less than the average earnings, mean = £43.4. We can conclude from this that there are a number of relatively high earnings which will of course affect the mean but not the median. Despite these uses of the mode and median, the mean is the most frequently used measure of central tendency. It is the 'expected' value, takes into account every value in the distribution and has many useful statistical properties that are explained as the book progresses. If we wanted a single measure of the number of cars exported per week last year or the number of days of work coalminers missed through injury we would use the mean.

2.5 Measures of Spread
We now have measures of the point of greatest frequency, mode, the expected value, mean, and the value of the middle observation, median, all of which refer to single points or class intervals. However, these measures tell us nothing about the general shape of the distribution. In Figure 2.3 both distribution (a) and distribution (b) have the same mean, median and modal range, yet their shapes are quite different.

If for example these two distributions represented the costs of two different courses of action although the expected cost is the same in both cases, the actual cost could only be as high as £11 in case (a) whereas it might be £15 in case (b). We need to have more information than just the mean in order to decide which course of action to take.

A simple descriptive measure of these differences in shape is to describe how far the distribution is spread out. We already have one example of this, namely the range. In Figure 2.3(a) the range is £9—£11 whereas in Figure 2.3(b) it is £5—£15. A number of other measures are used in practice.

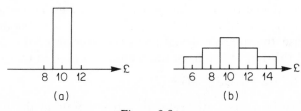

Figure 2.3

2.5.1 Quantiles

In the same way that the median divides the values of the distribution in half we can divide it into other proportions. For example we can estimate the value above which three-quarters of the population lie, and below which the remaining quarter lie. Similarly we could calculate the value below which three-quarters of the population lie, and above which a quarter lie, and so on for any other fraction we might choose. These points are known as 'quantiles'. The most frequently used quantiles divide the distribution into quarters and are called *quartiles* or into tenths when they are called *deciles*.

Quantiles are referred to by number from the lowest; thus the first decile has one-tenth of the population lying below it and nine-tenths above it; the second decile has two-tenths below it and eight-tenths above, and so on up to the ninth decile which has nine-tenths of the population below and only one-tenth above.

The quantiles therefore give us a further set of points indicating where the population lies. We can say directly that if an individual's income lies above a certain figure that he is in the top 10% of the income distribution. We can also look at the way the whole distribution has shifted between two time periods. While the median may have risen over time it is also possible for the bottom 20% of the population to have become poorer, or while the range has remained constant the quartiles may have moved closer to the median indicating a greater concentration or equality of incomes.

2.5.2 Interquartile Range and Quartile Deviation

Sometimes, it is useful to have a single measure of spread rather than a set of points, so that we can say simply that one distribution is more spread out than another. It is possible to obtain elementary measures from the quartiles. For example, where $Q1$ is the first quartile and $Q3$ the third quartile, the interquartile range is defined as $Q1$ to $Q3$ and the quartile deviation as $(Q3 - Q1)/2$.

Thus if the quartile deviation increases we can say that the middle 50% of the population is more spread out, and this would tend to imply that the whole distribution was more widely dispersed, although not necessarily so.

2.5.3 Variance and Standard Deviation

The quartile deviation has introduced a further concept into the analysis, that of the difference between two points in the distribution.

In this case the quartile deviation is the average of the distances of the first and third quartiles from the median, or second quartile.

$$(Q3 - Q1)/2 = [(Q3 - M) + (M - Q1)]/2$$

The quantiles, however, only refer to a few points in the distribution,

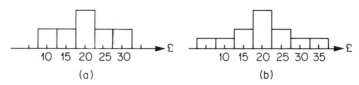

Figure 2.4

and their main convenience lies in the ease of their calculation as summary statistics. As is shown in Figure 2.4, two distributions with the same quartile deviations can have very different shapes, as the particular values of the lowest 25% and highest 25% of the population are not used in the calculation.

A more useful measure would make use of all the information about the values of the population, which we know. Each value, X, can be measured as a distance from a fixed point, a; so that we can talk about the deviation of X from a, $(X - a)$. We require some measure which will tell us that the distribution in Figure 2.4(b) is more spread out than the distribution in Figure 2.4(a). If we try summing $(X - a)$ then it is difficult to obtain much of any use to us. If $a = 0$ then $\Sigma(X - a) = \Sigma X$ and is the same in both cases. Similarly if we set $a = \mu$, $\Sigma(X - a) = 0$ for both cases. If we consider $\Sigma(X - a)^2$, however, these equalities no longer hold. The greater the spread the greater $\Sigma(X - a)^2$. For example if household expenditure were £4, £5 and £6 on the first three days of the week and £3, £5 and £7 on the second three days, $\Sigma(X - 0) = £15$ in both cases but $\Sigma(X - 0)^2 = £^2 77$ in the first case and $£^2 83$ in the second.

Our next problem is a sensible choice for a. If two distributions have exactly the same shape but the first has a larger mean than the second, then, if $a = 0$, ΣX^2 will be greater in the first case than the second. If $a = \mu$ on the other hand $\Sigma(X - a)^2$ will be the same for the two distributions. It is also the case that $\Sigma(X - a)^2$ is a minimum if $a = \mu$ which we can show quite easily.

★   (It is assumed that the reader can manage a little elementary
★   calculus at this stage. If this assumption is false, the reader should
★   try substituting alternative values for a in the right-hand side of (1)
★   to show that the expression increases in value the greater $|\mu - a|$,
★   where the vertical lines either side of an expression indicate that we
★   are to take the numerical value of the expression irrespective of
★   sign, e.g. if $\mu - a = -5$, $|\mu - a|$ is 5 just as it is if $\mu - a = 5$.)
★   Differentiating with respect to a
★
★ $$\frac{d\Sigma(X - a)^2}{da} = -2\Sigma(X - a) \tag{1}$$
★

★ Setting (1) equal zero to find a turning point

★

★ $$-2\Sigma(X-a) = 0$$

★ $$2\Sigma a = 2\Sigma X$$

★ $$\Sigma a = \Sigma X \tag{2}$$

★ Assuming that there are N members of the population we can
★ rewrite (2) as

★ $$Na = \Sigma X \tag{3}$$

★

★ and hence

★

★ $$a = \frac{\Sigma X}{N} = \mu \tag{4}$$

★

★ To check that we have a minimum,

★

★ $$\frac{d^2 \Sigma(X-a)^2}{da^2} = 2N \tag{5}$$

★

★ which is necessarily positive as N must be greater than zero.

We have introduced the concept of population size, N. Clearly if we consider $\Sigma(X-\mu)^2$ our measure will tend to increase with N. If our original population was size of pairs of socks produced by a machine, we could easily consider a new population of size of individual socks produced by the machine. The new distribution would have exactly the same shape as the original, but at every point along the range the frequency would be double. If we divide $\Sigma(X-\mu)^2$ by N, however, the two measures will be identical, and this measure

$$\frac{\Sigma(X-\mu)^2}{N}$$

is called the *variance* of the population, and is usually abbreviated as Var(X).

It is worth noting that Var(X) is an average of squared deviations round the mean, and that

$$\mathrm{Var}(X) = E[(X-\mu)^2] \tag{6}$$

There is one drawback in using the variance as a measure of spread in that it is measured in the square of the units of X. The concept of $£^2$ is not particularly attractive for example. If we take its square root the resulting measure is in the original units. This quantity

$$\sqrt{\frac{\Sigma(X-\mu)^2}{N}}$$

is called the *standard deviation* of the population and is usually abbreviated as σ_X.

2.5.4 Coefficient of Variation

Although in most cases we can be satisfied with the variance or standard deviation of the distribution as a measure of spread, we have not obtained a measure which is independent of the units of X. One solution to this which is sometimes used is to divide the standard deviation by the mean of the distribution. This dimensionless measure σ/μ is called the *coefficient of variation* and is usually expressed as a percentage, $100 \, \sigma/\mu$ %. However, it can only be used when all values of the population are positive, and is not invariant to a change in the origin. Thus if we are measuring the variance of a particular group of prices, while the coefficient of variation will remain the same if an *ad valorem* tax is imposed, it will change if a flat-rate tax is levied.

2.5.5 Measures of Dispersion Compared

As in the case of the measures of central tendency, the choice between measures of dispersion depends very much on the nature of the population and what we want to say about it. Quantiles provide further reference points in addition to the median whereas the standard deviation provides a single measure of spread which takes into account every member of the population. When we considered the two distributions in Figure 2.3 we suggested that a measure of spread was necessary in order to make a decision. A similar economic consideration can be made about stocks and shares. The decision about which assets to hold will be a function of the expected rate of return on each asset and the degree of risk involved. Thus if we look at the distribution of possible rates of return for an asset, their expectation is the mean and a measure of risk which is frequently used is their variance. In this particular case the difference in units between the two measures is not important as we are measuring two separate variables. If we are comparing, say, the distribution of profits in one industry with the distribution of profits in another where the mean values are, say, £4.2m and £4.5m respectively, and the standard deviations £3m in both cases, we would want to argue that the difference in means between the industries is small compared with the differences in profits of individual firms within each of the industries. If we had used the variances as our measure of spread such a direct comparison would not have been possible.

Finally, before proceeding to a consideration of the mechanics of calculating these measures of spread, we should bear in mind two points. Firstly the distributions we have considered have been symmetric either side of the mean. While we shall consider measures concerning lack of symmetry in Section 2.7, here we should merely note that asymmetry will also affect the usefulness of particular measures of spread. Secondly, much of the frequent use of the variance and standard deviation stems from their usefulness as statistics when we are drawing samples from the populations as we shall see in Chapter 3.

2.6 Problems of Calculation

The descriptive statistics which we just outlined are all extremely straightforward to compute. However, it is worth noting an easy method for computing the variance of a population. In Table 2.6 we have set out the simple frequency distribution of 'Duration of marriages where divorce or separation was granted' in Scotland during 1965. The first column sets out the class intervals, the second their midpoints, X, and the third their frequency, f.

We could proceed laboriously by calculating μ as $\Sigma fX/\Sigma f$ from column (4) and hence compute $(X - \mu)$, column (6) and finally $\Sigma f(X - \mu)^2/\Sigma f$ from column (7). However there are less tedious ways.

2.6.1 Using a Calculating Machine

We can simplify the variance as follows

$$\mathrm{Var}(X) = \frac{\Sigma f(X - \mu)^2}{\Sigma f}$$

$$= \frac{\Sigma fX^2}{\Sigma f} - \frac{2\Sigma fX\mu}{\Sigma f} + \frac{\Sigma f\mu^2}{\Sigma f}$$

$$= \frac{\Sigma fX^2}{\Sigma f} - \frac{2\Sigma fX\mu}{\Sigma f} + \mu^2 \tag{7}$$

Since

$$\Sigma fX/\Sigma f = \mu \tag{8}$$

we can substitute for (8) in (7)

$$\frac{\Sigma fX^2}{\Sigma f} - \frac{2\Sigma fX\mu}{\Sigma f} + \mu^2 = \frac{\Sigma fX^2}{\Sigma f} - 2\mu^2 + \mu^2$$

$$= \frac{\Sigma fX^2}{\Sigma f} - \mu^2 \tag{9}$$

We thus only need to calculate Σf, ΣfX and ΣfX^2 for each X, and these are set out in columns (3), (4) and (5) of Table 2.6. It is possible to calculate the sums of two of these columns simultaneously on many machines. In the case where we are using all the individual values of the population and not a grouped frequency distribution we can calculate ΣX and ΣX^2 simultaneously since $\mathrm{Var}(X) = \Sigma X^2/N - (\Sigma X/N)^2$.

Checking that the two methods are identical, we have $\Sigma f = 2,664$,

Table 2.6
Duration of marriages (where divorce or separation granted, Scotland, 1965)

(1) Duration[a]	(2) Midpoint X	(3) number f	(4) fX	(5) fX^2	(6) $(X-\mu)$	(7) $f(X-\mu)^2$	(8) $(X-a)^c$	(9) $f(X-a)$	(10) $f(X-a)^2$
Under 1 yr	0.5	—	—	—	—	—	—	—	—
1—2 yrs	1.5	21	31.5	47.25	−11.266	2,665.38	−13.5	−283.5	3,827.25
2—5 yrs	3.5	354	1,239	4,336.5	−9.266	30,394.00	−11.5	−4,071	46,816.5
5—10 yrs	7.5	975	7,312.5	54,843.75	−5.266	27,037.49	−7.5	−7,312.5	54,843.75
10—20 yrs	15.0	933	13,995	209,925	2.234	4,656.38	0	0	0
20 yrs and over	30.0[b]	381	11,430	342,900	17.234	113,161.10	15.0	5,715	85,725
Σ		2,664	34,008	612,052.5		177,914.35		−5,952	191,212.5
$\Sigma/\Sigma f$			12.766	229.75		66.78		2.2342	71.776

[a] Interpreting the class intervals as 1 year and less than 2 years etc.
[b] This is an assumption the reader may wish to challenge.
[c] $a = 15$ years.

Source: *Annual Abstract of Statistics*

$\Sigma fX = 34{,}008$ yrs, $\Sigma fX^2 = 612{,}052.5$ yrs^2. Thus using (9)

$$\text{Var}(X) = \frac{612{,}052.5}{2{,}664} - \left(\frac{34{,}008}{2{,}664}\right)^2$$

$$= 229.75 - 12.766^2 = 66.78$$

Using the long method

$$\text{Var}(X) = \frac{177{,}914.35}{2{,}664} = 66.78$$

2.6.2 By Hand

In this case we wish to simplify the arithmetic as far as possible without losing accuracy. It is therefore desirable to keep the number of significant digits as small as possible and the numbers themselves simple.

Firstly we should try to simplify X, column (2). We may be able to remove a common factor, say by dividing by 5 or 100, although this will not help in our particular example. However, we can cut down the absolute size of the numbers without increasing the number of digits if we compute $(X - a)$ instead of X, where a is an arbitrary constant, usually equal to that value of X which we *guess* before calculating is nearest to the mean. In this case either 7.5 of 15.0 is likely to be closest to the mean, so we could take either, but we have chosen 15 as an illustration. We can thus calculate the variance round a instead of μ as in column (10), $\Sigma f(X - a)^2 / \Sigma f$. We must then subtract the difference between this and $\text{Var}(X)$, which is

$$\frac{\Sigma f(X - a)^2}{\Sigma f} - \frac{\Sigma f(X - \mu)^2}{\Sigma f} = \frac{\Sigma fX^2}{\Sigma f} - \frac{2\Sigma fXa}{\Sigma f} + a^2 - \frac{\Sigma fX^2}{\Sigma f} + \left(\frac{\Sigma fX}{\Sigma f}\right)^2$$

$$= \left(\frac{\Sigma fX}{\Sigma f}\right)^2 - 2\frac{\Sigma fX}{\Sigma f}a + a^2$$

$$= \left(\frac{\Sigma fX}{\Sigma f} - a\right)^2 = (\mu - a)^2 \tag{10}$$

Hence

$$\text{Var}(X) = \frac{\Sigma f(X - a)^2}{\Sigma f} - \left(\frac{\Sigma fX}{\Sigma f} - a\right)^2 \tag{11}$$

It is easy to see that the calculating machine technique uses (11) with $a = 0$. From columns (9) and (10)

$$\text{Var}(X) = \frac{191{,}212.5}{2{,}664} - \left(\frac{-5{,}952}{2{,}664}\right)^2$$

$$= 71.776 - (-2.2342)^2 = 66.78$$

2.7 Skewness

All the distributions we have considered in Figures 2.3 and 2.4 are symmetric about their means. In other words their shape is the same on both sides of the mean; the frequency at $(\mu - a)$ is equal to that at $(\mu + a)$, where a is an arbitrary constant. Yet in all the examples we have shown from real life the distributions are not symmetric, they are skewed. The mode and mean of the distributions do not coincide, and more importantly the mean and median do not coincide either. If values are concentrated towards the lower end of the range, as in Figure 2.5(a), the distribution is *positively* skewed, and if they are concentrated towards the upper end they are negatively skewed, as in Figure 2.5(b).

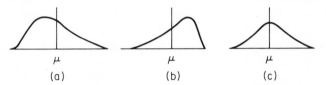

$$\mu \qquad \mu \qquad \mu$$
$$(a) \qquad (b) \qquad (c)$$

Figure 2.5 (a) Positive skewness. (b) Negative skewness. (c) Symmetrical

A simple measure of skewness can be obtained from the quartiles.

$$\frac{(Q3 - M) - (M - Q1)}{(Q3 - M) + (M - Q1)}$$

is called the *quartile measure of skewness*. If the distribution is symmetric then $(Q3 - M) = (M - Q1)$ and the skewness is zero. The maximum degree of skewness possible is that the first or third quartile should equal the median. For example, if we find out how many cars each person in a group owns, it is likely that $Q1 = 0$ and $Q3 = M = 1$. In this case the skewness is -1, maximum negative skew. Similarly maximum positive skew is $+1$.

Let us take the distribution of net wealth from Table 2.2 as an example of skewness. Firstly we must calculate the three quartiles. The total frequency is 19,483,000, but let us conduct the arithmetic in thousands of individuals to keep it simple. The median, lower and upper quartile observations are thus the

$$\frac{19,483}{2} = 9,742 \qquad \text{(thousandth)}$$

$$\frac{19,483}{4} = 4,871 \qquad \text{(thousandth)}$$

$$\frac{19,483}{4} \times 3 = 14,612 \qquad \text{(thousandth)}$$

respectively (to the nearest thousand). From the cumulative frequencies shown in Table 2.2 we can see that $Q1$ lies in the interval 'less than £1000', M in '£1,000—£3,000' and $Q3$ in '£3,000—£5,000'. Evaluating

$$Q1 = £1,000 \times (4,871/8,671) \qquad\qquad = £562$$
$$M = £1,000 + £2,000 \times (9,742 - 8,671)/5,582 \quad = £1,384$$
$$Q3 = £3,000 + £2,000 \times (14,612 - 14,253)/2,361 \quad = £3,304$$
$$QMS = \frac{(3,304 - 1,384) - (1,384 - 562)}{(3,304 - 1,384) + (1,384 - 562)} = \frac{1,098}{2,742} \quad = 0.400$$

The distribution of net wealth is thus positively skewed, which is what we would expect, since most of the wealth is concentrated in a fairly small number of hands. In fact if we had a measure which included all the members of the population, we would find that the distribution was even more positively skewed. Such a measure can be obtained in a like way to the standard deviation, but that lies outside our present purpose.

2.8 Theoretical Distributions
We now have a number of statistics to help us describe distributions. These do not represent the total number which can be used helpfully, but they are those in common use. The next step would be to consider kurtosis, the peakedness of the distribution. We should also note that some of these statistics can have more than a descriptive role. These measures of the characteristics of the population are referred to as parameters.

If we can express the shape of a distribution in some functional form, we can use these parameters to determine the exact size of the distribution. Up till now most of the distributions we have dealt with have been either discrete or continuous approximations to discrete distributions. The two theoretical distributions we shall deal with are continuous, and thus they have a frequency and cumulative frequency for every point on the range. These two distributions have been drawn in Figure 2.6. The first of these is perhaps the simplest of all

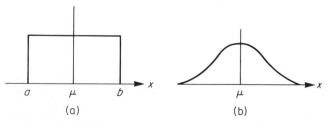

Figure 2.6 (a) Rectangular distribution. (b) Normal distribution. Frequency is shown by the vertical axis

distributions; it has a single uniform frequency, f, over the range a to b, and is known as the rectangular or uniform distribution. The second, which forms a fundamental concept in statistics is known as the normal distribution, but the function of its frequency is more difficult to express. However, let us leave the normal distribution for the moment.

We have already used the rectangular distribution on a number of occasions. In drawing histograms and finding midpoints we assumed that the values were rectangularly distributed across each class interval. Let us pursue its properties a little further. Since there is a frequency for every value of a variable, x, in a continuous distribution we can express frequency as a function of x. We shall call $f(x)$ the *frequency function* of the distribution. Thus for the rectangular distribution

$$f(x) = k \quad a < x < b \tag{12}$$
$$= 0 \quad \text{otherwise}$$

where k is a constant. It will also be convenient if we assume that we are measuring relative frequencies, $f/\Sigma f$, rather than the absolute frequencies. Thus in the discrete case $\Sigma f = 1$.

In the same way we can also derive a function for cumulative frequency. In the discrete case $F(x)$, the cumulative frequency function, is equal to the sum of the frequencies, $f(x)$, up to x,

$$F(x) = \sum_{i}^{x} f(x_i) \tag{13}$$

In order to look at the continuous case we must make use of a little calculus. For those who are not conversant with the calculus it will be sufficient to accept that $f(x)$ gives the functional form of the frequency curve and $F(x)$ gives the area under the frequency curve. Thus from Table 2.6, when $x = 3.5$, $f(x) = 354/2,664$ and $F(5) = 375/2,664$ (i.e. $f(1.5) + f(3.5) = (21 + 354)/2,664$). We shall call $F(x)$ the *distribution function*.

★ In the continuous case
★
★
★ $$F(x) = \int_{-\infty}^{x} f(x)\mathrm{d}x \tag{14}$$
★
★ and as the result of our assumption of relative frequency, the total
★ frequency is unity.
★
★ $$F(\infty) = \int_{-\infty}^{\infty} f(x)\mathrm{d}x = 1 \tag{15}$$
★
★ In the case of the rectangular distribution in Figure 2.6(a)
★
★ $$f(x) = 1/(b-a) \quad a < x < b \tag{16}$$
★ $$= 0 \quad \text{elsewhere}$$

★ and

$$F(x) = \int_a^x f(x)\mathrm{d}x = \int_a^x 1/(b-a)\mathrm{d}x \qquad (17)$$

hence evaluating

$$F(b) - F(a) = \int_a^b 1/(b-a)\mathrm{d}x$$

$$= [x/(b-a)]_a^b$$

$$= b/(b-a) - a/(b-a)$$

$$= 1 \qquad (18)$$

The normal distribution has the following form

$$f(x) = \frac{1}{\sigma\sqrt{(2\pi)}} \left\{ \exp\left[\frac{-(x-\mu)^2}{2\sigma^2} \right] \right\} \qquad (19)$$

(where exp is short for exponent and means that the expression which follows is a power of e; e.g. $\exp(a) = e^a$). Hence

$$F(x) = \frac{1}{\sigma\sqrt{(2\pi)}} \int_{-\infty}^x \left\{ \exp\left[\frac{-(x-\mu)^2}{2\sigma^2} \right] \mathrm{d}x \right\} \qquad (20)$$

Again $F(\infty) = 1$.

These expressions are complex, and we shall not explain them here. In the next chapter we shall see that the normal distribution is fundamental to the subject of sampling. However, at this stage it is worth pointing out that many empirical populations are approximately normally distributed. If we take a machine which is set to produce objects of a fixed size, but which is subject to vibration, we may well find that it tends to produce objects whose size is normally distributed round the mean of the set size. Vibration and other disturbances tend to put the machine off, but the greater the difference between a size and the mean the less frequently it is produced by the machine. Figure 2.6(b) is often describes as 'bell-shaped'. The nearer the mean the greater the frequency, but however far away from the mean a value is it always has a positive frequency. The frequency function is only zero at $-\infty$ and $+\infty$.

There are only two variable elements in the normal distribution function, μ and σ^2. These two are the *parameters* of the distribution. If we wish to refer to a particular normal distribution we express it as $N(\mu,\sigma^2)$. We shall make a lot of use of the distribution which is $N(0,1)$. It is obvious that as μ varies the distribution will be shifted along the range. Similarly as σ^2 increases the distribution becomes flatter and more spread out.

★ (While we are considering continuous distributions it is also worth
★ noting that while $E[X] = \Sigma f x$ in the discrete case, in the continuous
★ case $E[X] = \int x f(x)\mathrm{d}x$, and similarly $E[X^2] = \int x^2 f(x)\mathrm{d}x$.)

References and Suggested Reading

1. An excellent introduction to statistics and frequency distributions is given in

M. G. Kendall and A. Stuart, *The Advanced Theory of Statistics*, 3rd Edition, Charles Griffin & Co., London, 1969, Vol. 1, Ch. 1, pp. 1—15.

The reader should not be put off by the title of the book or by a look at the rest of the material in it. Those who wish to avoid the calculus should confine themselves to pp. 1—11 only.

2. There are several books providing an introduction to mathematics for economists. Most of these cover far more material than will be useful here. Those interested in the calculus might like to look at

J. Black and J. F. Bradley, *Essential Mathematics for Economists*, John Wiley & Sons, London, 1973, Chs. 5, 9 and 6 (in that order).
J. Parry Lewis, *An Introduction to Mathematics for Students of Economics*, Macmillan, London, 1965.
R. G. D. Allen, *Mathematical Analysis for Economists*, Macmillan, London, 1966.

3. In this chapter we introduced a major source of annual data concerning the U.K., which the reader will find very useful in the study of economics.

Annual Abstract of Statistics, H.M.S.O., London, annually.

Questions

1 Using the data given below, form a frequency distribution of the average monthly Treasury Bill rates during the period 1962 to 1964. Draw a histogram of the distribution, and comment on its shape.

Average Treasury Bill rates(%)

Month	1962	1963	1964
January	5.352	3.511	3.724
February	5.417	3.454	3.918
March	4.862	3.555	4.301
April	4.258	3.711	4.301
May	3.935	3.672	4.346
June	3.803	3.693	4.442
July	3.902	3.768	4.569
August	3.783	3.708	4.654
September	3.688	3.691	4.653
October	3.709	3.673	4.689
November	3.773	3.753	5.107
December	3.664	3.738	6.623

Source: *Annual Abstract of Statistics*

2   Compare the distribution of the U.K.'s imports by source (country) in 1961 with that in 1969 from the data below. Use the medians, quartiles, interquartile ranges and quartile measures of skewness in your comparison.

Value of imports (analysis by source)

	Frequency	
Value	1961	1969
Less than £50 mn	16	7
£50 mn and less than £100 mn	10	9
£100 mn and less than £150 mn	5	6
£150 mn and less than £200 mn	8	6
£200 mn and less than £500 mn	2	11
Greater than £500 mn	0	2

Adapted from *Annual Abstract of Statistics*

3   Calculate the median output of the car manufacturer shown below. Is a frequency polygon a useful method of showing the shape of this distribution?

Daily output of completed cars

Day	1	2	3	4	5	6	7	8	9	10	11	12	13	14
Output	3	7	12	5	18	6	2	21	13	4	15	16	17	8

4   Estimate the mean, variance and standard deviation of coal production in the areas shown for 1969/70. Compare the spread of this population with that of output per manshift for the same areas in 1969/70.

Coal production and output per manshift, 1969/70

Area	Production (mn tons)	Output per manshift (cwt)
Scottish North	4.98	38.5
Scottish South	6.39	35.6
Northumberland	6.85	42.4
North Durham	5.36	33.4
South Durham	8.29	36.3
North Yorkshire	9.22	49.4
Doncaster	8.53	47.9
Barnsley	8.07	42.0
South Yorkshire	10.10	46.2
North Western	7.01	36.0
North Derbyshire	10.59	56.0
North Nottingham	11.78	58.7
South Nottingham	11.15	58.6
South Midlands	9.26	59.7
Staffordshire	8.56	50.5
East Wales	7.72	27.2
West Wales	5.07	29.7
Kent	1.10	26.6

Source: *Annual Abstract of Statistics*

Answers

1 We must firstly decide on a set of class intervals. As there are only 36 members of the population, 6—8 intervals will be sufficient to give us meaningful frequencies. Most of the observations lie in the range 3.5—4.5%, therefore, the minimum interval must clearly be less than 1%. We have set out a possible choice below.

Treasury Bill rate, 1962—64

%	Frequency	Frequency per 0.25%	Tally
Less than 3.5	1	1	/
3.5 and less than 3.75	13	13	‖‖‖ ‖‖‖ ///
3.75 and less than 4.0	8	8	‖‖‖ ///
4.0 and less than 4.25	0	0	
4.25 and less than 4.5	5	5	‖‖‖
4.5 and less than 5.0	5	2.5	‖‖‖
Greater than 5.0	4	0.5	////

Source: *Annual Abstract of Statistics*

On the right-hand side of the table we have shown how a simple set of tally marks can be used in calculating frequencies. The second column shows the frequencies calculated.

Before constructing the histogram we must estimate frequencies for a common interval. The simplest choice is the minimum interval, 0.25%, and the results of this are shown in the third column. We have two open-ended classes which we can only deal with by assumption. We have closed them at 3.25% and 7% respectively, although we do in fact know the individual values. We can now draw the histogram (Figure 2.7) using column 3 as the vertical axis, so that the frequency per class interval will be shown by area. (We have indicated by the dotted line, — — —, the effect on the histogram of amalgamating the 4th and 5th class intervals.)

There are a number of general comments that we can make about this distribution. Firstly although the range is 3.25% to 7% the bulk of

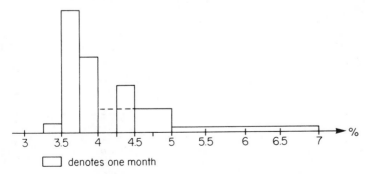

Figure 2.7   Histogram of Treasury Bill Rate, 1962—64 (monthly averages)

the values lie in the range 3.5% to 4%. The modal range is 3.5% to 3.75% although there is a minor modal range of 4 25% to 4.5%. Secondly the use of a histogram to examine these data precludes a study of their most interesting aspect, which is the change in the values over time. It is debatable whether we should amalgamate the 4th and 5th class intervals of the range 4% to 4.5% in order to make the histogram smoother, since there appear to be two groups of rates, the first and largest being in the range 3.5% to 4% and the second 4.25% to 5%, whose distinction should be brought out. However, we should be wary of creating non-existent features for a population purely because of our choice of class intervals.

2   Let us firstly calculate the required statistics.

Value of imports (analysis by source)

Value £mn	1961 f	1969 f	1961 F	1969 F
Less than 50	16	7	16	7
50 and less than 100	10	9	26	16
100 and less than 150	5	6	31	22
150 and less than 200	8	6	39	28
200 and less than 500	2	11	41	39
Greater than and equal to 500	0	2	41	41
Total	41	41		

$Q1$ is the value of the $\Sigma f/4 = 10.25$th 'observation'.
M is the value of the $\Sigma f/2 = 20.5$th 'observation'.
$Q3$ is the value of the $3\Sigma f/4 = 30.75$th 'observation'.
We can calculate these using the following generalized procedure:

(a) find the class interval in which Q lies

(b) $Q = L + W \left(\dfrac{G - F}{f_Q} \right)$

where L is the value of the lower bound of the class,
  W is the class width,
  G is the number of the quartile observation,
  F is the cumulative number of observations in all previous classes,
  f_Q is the number of observations in the class.

1961	1969
$Q1$ is in the first interval	$Q1$ is in the second interval
Assumption: interval begins at zero	
$Q1 = 50 \times 10.25/16$ = £32.03 mn	$Q1 = 50 + 50 \times (10.25 - 7)/9$ = £68.06 mn
M is in the second interval	M is in the third interval
$M = 50 + 50 \times (20.5 - 16)/10$ = £72.50 mn	$M = 100 + 50 \times (20.5 - 16)/6$ = £137.50 mn

1961	1969
$Q3$ is in the third interval	$Q3$ is in the fifth interval
$Q3 = 100 + 50 \times (30.75 - 26)/5$	$Q3 = 200 + 300 \times (30.75 - 28)/11$
$= £147.50$ mn	$= £275.00$ mn
Interquartile range = £115.47 mn	Interquartile range = £206.94 mn
from £32.03 mn to £147.50 mn	from £68.06 mn to £275.00 mn
Quartile measure of skewness	

$$= \frac{(Q3 - M) - (M - Q1)}{(Q3 - M) + (M - Q1)}$$

QMS

$$= \frac{(147.50 - 72.50) - (72.50 - 32.03)}{(147.50 - 72.50) + (72.50 - 32.03)}$$

$$= \frac{(275.00 - 137.50) - (137.50 - 68.06)}{(275.00 - 137.50) + (137.50 - 68.06)}$$

$= 0.30$ 　　　　　　　　　　　　 $= 0.33$

Using these descriptive statistics we can now compare the two distributions. We should note the following points

(a) In 1969 all three quartiles are larger than in 1961 and are virtually doubled in size.

(b) Since we are dealing in £mn we would expect (a) from inflation alone.

(c) Both distributions have a similar positive skew — indicating that we have a small amount of trade with many countries, but a relatively much larger amount with a few countries. (This is a function of size of exporting country as well as type of products and geographical nearness to the U.K.)

(d) Since our statistics only deal with the middle 50% of the population they do not take into account the considerable reduction in skewness in the 1969 case.

(e) The open-ended class of greater than £500 mn means that we cannot establish the full extent of the increase in range of the 1969 values.

3　Firstly we must rearrange the population in order of magnitude.

Daily output of completed cars

2	3	4	5	6	7	8	12	13	15	16	17	18	21

There are 14 members of the population, therefore, the median is the average of the 7th and 8th observations

$$M = \frac{8 + 12}{2} = 10 \text{ cars}$$

Since we have each member of the population identified we do not have a frequency distribution, so a frequency polygon is an inappropriate descriptive method. With only 14 observations there is little point in creating such a frequency distribution. Lastly the distribution is discrete, so the use of a continuous line for frequency would not be of any real meaning.

4   Coal production and output per manshift

	Production (mn tons)		Output/manshift (cwt)	
	X	X^2	Y	Y^2
	4.98	24.80	38.5	1,482.25
	6.39	40.83	35.6	1,267.36
	6.85	46.92	42.4	1,797.76
	5.36	28.73	33.4	1,115.56
	8.29	68.72	36.3	1,317.69
	9.22	85.01	49.4	2,440.36
	8.53	72.76	47.9	2,294.41
	8.07	65.12	42.0	1,764.00
	10.10	102.01	46.2	2,134.44
	7.01	49.14	36.0	1,296.00
	10.59	112.15	56.0	3,136.00
	11.78	138.77	58.7	3,445.69
	11.15	124.32	58.6	3,433.96
	9.26	85.75	59.7	3,564.09
	8.56	73.27	50.5	2,550.25
	7.72	59.60	27.2	739.84
	5.07	25.70	29.7	882.09
	1.10	1.21	26.6	707.56
Σ	140.03	1,204.81	774.7	35,369.30
Σ/N	7.779	66.93	43.04	1,964.96 ($N = 18$)

$\mu_X = 7.779$ $\mu_Y = 43.04$

$$\text{Var}(X) = \frac{\Sigma X^2}{N} - \mu_X^2$$ $\text{Var}(Y) = 1,964.96 - 43.04^2$

$$= 66.93 - 7.779^2$$ $= 112.5$

$$= 6.42$$

$\sigma_X = 2.53 = \sqrt{\text{Var}(X)}$ $\sigma_Y = 10.61$

Coeff. of variation $= \dfrac{100\sigma\%}{\mu}$ Coeff. of variation $= 24.7\%$

$$= 32.5\%$$

Comparison of the two standard deviations would not be very enlightening owing to the different units of measurement. We can see from the coefficients of variation that output/manshift is relatively less dispersed than production in the various regions.

3 Probability and Sampling

Probability

3.1 The Concept of Probability

Probability is a concept with which we are all familiar. If we decide before going out that it will 'probably' rain then we would take an umbrella or a raincoat. If we think that share prices are rising then it would be rational to buy now rather than later. In both these cases we are assigning some implicit value to the 'likelihood' of these events occurring and our subsequent action then depends on what value we place on various levels of probability — a decision based, perhaps, on past experience.

The probability of an event, say A, occurring is the number of times A occurs divided by the total number of trials held. We have to imagine ourselves conducting what we call a random experiment, by which we mean that a particular outcome cannot be predicted. The term experiment is used in statistics to describe any act which is repeatable under the same conditions. The number of times a random experiment is performed constitutes the number of trials. In economics we are usually presented with the results of natural random experiments; we cannot, under most circumstances, experiment with the economy as a whole, or even with individual people's behaviour. However, some experimental situations occur in the course of normal human activity; we could for example set up a random experiment by noting the sex of the next person to walk down the street. If asked to give the probability that this person would be male, most people would say that they expected there to be an equal number of males and females, so that the ratio of males to the total population would be one-half. We could calculate a more precise value by consulting official publications giving population figures, such as Table 6 in the *Annual Abstract of*

Statistics. This is an *a priori* way of looking at probability which can be defined in the following way:

Suppose that a trial can end in N different, equi-likely ways, M of which give rise to an event A. We define the probability of A as P(A) = M/N.

M must obviously be greater than or equal ($\geqslant$) to 0 and less than or equal to N, i.e. $0 \leqslant M \leqslant N$, giving $0 \leqslant P(A) \leqslant 1$.

0 and 1 thus form the limits to our probability range. It is generally agreed that the lower limit 0 is the probability attached to the occurrence of an impossible event. The upper limit, 1, is the probability attached to the occurrence of a certain event. For example,

P(death for an individual) = 1

P(throwing a 7 with an ordinary die) = 0

If we use past experience to help us evaluate the probability of an event, then we use an *a posteriori* or relative frequency definition and say that:

If a trial is repeated *n* times, *m* of which produce the event *A*, then

$$P(A) = \lim_{n \to \infty} \frac{m}{n}$$

★ lim stands for 'in the limit' and $n \to \infty$ tells us that the limit is
★ reached as *n* tends to infinity.

The ratio m/n, the relative frequency, can be empirically observed to become stable in the long run, i.e. as *n* gets very large. The commonest example of this is in tossing a fair coin. There are two possible outcomes, a head or a tail. If the coin is tossed ten times then perhaps eight of the outcomes will be a head. If, however, we repeat this experiment a greater number of times then we will find that the probability of getting a head will approach 0.5. The larger *n* the nearer P(a head) is to 0.5.

We would use the relative frequency approach if we wished to establish the probability that a customer in a particular large super-market includes butter among his purchases. We cannot establish an *a priori* probability because although we know that most people eat butter, we have only the vaguest idea about the proportion of occasions on which they purchase it when they visit a supermarket. Thus if after a month of recording purchases, we found that on 37% of occasions a customer had butter among his purchases, we would say that the probability of a customer purchasing butter on any one occasion was 0.37.

3.2 Simple Events

The events we have been considering have been simple events. A *simple* (or elementary) *event* is an outcome of an experiment that consists only of single events and not of any combinations of events. Each of the outcomes of tossing a coin is a simple event. However if we combine two or more simple events we obtain a *compound* (or composite) *event*. We will look at these in more detail shortly but first we must return to simple events.

Suppose an experiment which is repeated n times, where n is a large number, has A_i ($i = 1,2, \ldots ,k$) outcomes. When the experiment is performed one and only one of these outcomes is observed, i.e. we are talking about a simple event. Let m_i ($i = 1,2, \ldots , k$) be the number of times outcome A_i occurs. Then because one and only one outcome occurs on each trial of the experiment

$$m_1 + m_2 + \ldots + m_k = n$$

Dividing by n gives:

$$\frac{m_1}{n} + \frac{m_2}{n} + \ldots + \frac{m_k}{n} = 1$$

Since n is large the m_i/n approach $P(A_i)$ so we have that

$$\sum_{i=1}^{k} P(A_i) = 1 \tag{1}$$

and remember that

$$0 \leqslant P(A_i) \leqslant 1 \tag{2}$$

3.3 Compound Events

We have said that a compound event consists of two or more simple events. Let us take an example; the event set of the outcomes of tossing two coins can be shown as follows:

		Second coin	
		H	T
First	H	HH	HT
coin	T	TH	TT

giving four simple events. If we now take our event A as being that where we have just one tail (and thus just one head) then we see that A is the compound event consisting of the two simple events HT and TH. The probability of any one of the four simple events is ¼. The question arises of the probability of our compound event and we define this to be equal to the sum of the probabilities of the composite simple events. In this example this gives us a probability of ¼ + ¼ = ½.

3.4 Addition and Multiplication Rules of Probability

(1) Addition Rule We need first to define the term '*mutually exclusive*':

Two events, A and B, are said to be mutually exclusive if no simple event belongs to both A and B.

We can now say that:

If A and B are two mutually exclusive events then the probability of A or B occurring (includes both occurring) in any given trial is equal to the sum of the individual probabilities of A and B occurring. I.e.

$$P(A \text{ or } B) = P(A) + P(B) \tag{3}$$

★ Those familiar with set theory notation will recognize this as
★ $P(A \cup B) = P(A) + P(B)$. Using set theory to introduce the concept
★ of probability is a common approach, but is not a central enough
★ concept for this book and is therefore not included. Set theory is
★ quite easy to understand and is adequately covered in many
★ elementary statistical books, for example R. E. Walpole, *Intro-*
★ *duction to Statistics*, 2nd Edition, Collier—Macmillan, 1974.

If, though, the events A and B are not mutually exclusive then we have:

$$P(A \text{ or } B) = P(A) + P(B) - P(A \text{ and } B) \tag{4}$$

★ i.e.
★
★ $$P(A \cup B) = P(A) + P(B) - P(A \cap B)$$

This can easily be seen if we take as A the event that a householder owns a car and B as the event that a householder owns a washing machine. Obviously not everyone who owns a car will also own a washing machine but there will be some people in this category. If therefore we were to use equation (3) and just add the probabilities we would have included some people twice, so we have to subtract the probability that A and B both occur. We could use equation (3) if we were to take A as the event that a man is driving his own car and B as the event that he is driving his wife's car. He obviously cannot drive both at the same time and so the events are mutually exclusive.

(ii) The Multiplication Rule This introduces another concept — that of *conditional probability*. This is defined to be:

The probability that B occurs given the occurrence of A. It is written as $P(B|A)$.

Thus given a group of 20 people and the information that 16 of them held a driving licence and 15 of these owned a car, the probability that anyone picked at random from the group of 20 would be a car owner would be 15/20. If, though, we were told that our randomly picked person held a driving licence then with this additional information we could say that the probability that this person owned a car was 15/16, i.e.

P(person owned car | person held driving licence) = 15/16

We are now in a position to state the multiplication rule of probability:

$$P(A \text{ and } B) = P(A)P(B \mid A) = P(B)P(A \mid B) \tag{5}$$

★ or
★ $\qquad P(A \cap B) = P(A)P(B \mid A)$

I.e. the probability of both A and B happening is equal to both the probability that A happens multiplied by the conditional probability that B happens given that A happens, and also to the probability that B happens multiplied by the conditional probability that A happens given that B happens.

Using the data of the previous example we have that the probability that a person picked at random from a group of 20 is both a licence holder and a car owner is

$$\frac{16}{20} \cdot \frac{15}{16} = \frac{3}{4}$$

or

$$\frac{15}{20} \cdot 1 = \frac{3}{4}$$

If the two events A and B are *independent*, i.e. the occurrence or non-occurrence of one has no effect on the occurrence of the other, then obviously

$$P(B \mid A) = P(B) \quad \text{and} \quad P(A \mid B) = P(A)$$

equation (5) then becomes

$$P(A \text{ and } B) = P(A)P(B) \tag{6}$$

Results (3), (4), (5) and (6) may all be extended for further events.

3.5 Complementary Events
Any event must either occur or not occur. From (1) and (3) we have that:

$$P(\text{event occurs}) + P(\text{event does not occur}) = 1$$

If A is our event, then P(event does not occur) is written as $P(\bar{A})$ and $\bar{A}$ is known as the *complement* of A. Therefore

$$P(A) = 1 - P(\bar{A})$$

This is a very useful relationship since it is often the case that $\bar{A}$ is easier to find than A. For example, the probability that an individual has owned more than one car in their lifetime would be more easily found as:

$$P(\text{owned more than one car}) = 1 - [P(\text{owned 0 car}) + P(\text{owned 1 car})]$$

than as

$$P(\text{owned 2 cars}) + P(\text{owned 3 cars}) + \ldots$$

3.6 Permutations and Combinations

Many experiments will have a large number of possible outcomes. It is not always possible to list these, and in any case it is rarely advisable to try, and so we resort to the techniques of permutations and combinations. These provide a convenient way of counting.

An arrangement of objects in a definite order is called a *permutation*. For example, the number of possible permutations of the three letters A, B and C is 6; ABC, BCA, CAB, ACB, BAC, CBA. There are 3 ways of filling the first place, this leaves two ways of filling the second and one for the last, $3 \cdot 2 \cdot 1 = 6$.

In general, n distinct objects can be arranged in $n(n-1)$ $(n-2) \ldots 3 \cdot 2 \cdot 1$ ways. We denote this product by $n!$ and call it n factorial. We say that the number of permutations of n different objects taken all together is $n!$ and denote it by ${}_nP_n$. So

$$ {}_nP_n = n!$$

If we have only r spaces for our n objects, $r \leqslant n$, then we have the number of permutations of n different objects taken r at a time as

$$ {}_nP_r = n(n-1)(n-2) \ldots (n-r+1)$$

$$ = \frac{n!}{(n-r)!} \qquad r \leqslant n$$

For the case where $r = n$ we use the fact that by definition $0! = 1$, e.g. if $n = r = 2$,

$$ {}_2P_2 = \frac{2!}{(2-2)!} = \frac{2!}{0!} = \frac{2}{1} = 2$$

If we select objects without regard to order then we are using *combinations*. For example, how many ways can you arrange 3 letters taken 2 at a time? If order is taken into account then there are

$_3P_2 = 3!/(3-2)! = 6$ different ways, AB, BC, CA, AC, BA, CB. This though is implying that AB is different from BA; if order is ignored there are 3 ways.

In general the number of combinations of n different objects taken r at a time is

$$_nC_r = \frac{n!}{r!(n-r)!}$$

Thus

$$_nC_r = {_nP_r}\,\frac{1}{r!}$$

the factor $1/r!$ appears because a set of r objects gives rise to $r!$ permutations and we wish to select without taking the order of the objects into consideration.

SAMPLING

3.7 The Taking of Samples

Being of a conservative nature most of us practise sampling fairly frequently — we take just a little to see if it is all right. A sip of soup in case it is too hot, one toe in the bath in case the water is too hot, a taste of wine before accepting it. Statistical sampling is trying to achieve the same objectives — finding out what the characteristics are of a population by looking at some of its members. Obviously if we want complete and accurate information about a particular characteristic of a population then we should use every member of that population. It is immediately apparent though that although in many cases this will be a long and tedious task, in others it will be well nigh impossible. We may also find that a complete enumeration is less accurate due to mistakes made than a well designed sample.

The main reasons for sampling may be summarized as follows:

(i) *Practicability*. A population may be either large or infinite and a complete census impossible.

(ii) *Cost*. The cost in time and money may well be prohibitive for a full census.

(iii) *Feasibility*. Where a product is to be tested for durability then it is obviously not a good idea to test every item — there would be none left to use or sell.

If we are to use the results of our samples to generalize about the populations they come from, i.e. to use sample information to estimate population parameters, then we must be sure that our sample is truly representative of the parent population and that it is not biased in any way. Our best way of tackling this problem is to use a random sample.

This does not describe the sample design, but rather the way in which the sample is picked. Picking things at random may initially sound like a very hit-and-miss way of proceeding, but to ensure that samples are picked at random demands a high degree of efficiency. We have to ensure that every member of the population has an equal probability of being picked for the sample and we can determine this probability for a random sample of size n drawn from a population of size N. There will be $_N C_n$ possible samples, all with an equal probability of being picked and the probability that any one of these random samples will be picked is $1/_N C_n$.

It is unfortunately not enough for us to devise our own method of 'random' selection; however random we think we are being we will undoubtedly introduce some bias into the selection process. We could stand outside a cinema and ask every tenth person entering how much they spent on entertainment. This could be faulted on two grounds: firstly people entering a cinema may spend more than average on entertainment, secondly, there may be something special about every tenth person, people might be entering in pairs, female first, so that every tenth person was male. The best way of achieving a random sample is to number each item in the population to be sampled and then draw the numbers by means of *random number tables*. These can be found in Appendix Tables A3.1 and consist of four digits, each an independent sample from a population in which the digits 0 to 9 have a probability of 1/10 of occurring. We can start anywhere in the table, say the top of column 4 and read off numbers that could correspond to our enumerated population. For example, if our population consisted of 400 firms then we would read off the first 3 digits, obtaining 255, 011, 143 etc.; and we would then pick these firms out for our sample. If our population was the 43 British Airports (see *Business Monitor, Civil Aviation Series*) then we would just take the first two digits every time until we had our sample. If the number drawn from the table does not correspond to any numbered item in our population, for example 44, we would merely pass on to the next number.

Assigning numbers to each member of our population and then including in the sample those members whose number is given by the random number tables is a method of random sampling which becomes less applicable as the population size increases. With a larger population we have to turn to other methods of random sampling.

An example of sampling from a very large population is given by the *Family Expenditure Survey*, (normally abbreviated as the F.E.S., an annual survey published by H.M.S.O.). The Department of Employment who conduct the survey are faced with the task of selecting about 11,000 addresses from all the households in Great Britain. The aim is to obtain information about the pattern of expenditure, incomes and the composition of private households. Results from the F.E.S. are used to weight the General Index of Retail Prices which replaced the Cost of Living Index, though the latter term is still in colloquial use. (For an explanation of weighting procedures see Appendix A of the F.E.S.) The General Index is being used increasingly in pay negotiations, though it can be safely said that the whole of the F.E.S. provides a fund of very important economic and social data.

The households selected for the survey obviously have to be representative and this is the rationale behind the rather complicated method of sampling used. We would draw some very peculiar inferences if all the members of the sample had low incomes and lived in a remote corner of Scotland. Clearly we must divide the population into a number of broad categories and draw random samples from each of these to make sure we have a sufficiently broad distribution of households to draw helpful conclusions. In the F.E.S. Great Britain is subdivided by type of area, region and an economic factor in the form of rateable values. In the final stage when the breakdown has gone as far as parishes and wards, random selection from the Registers of Electors gives the addresses to use.

★ This is called a three-stage stratified rotating design. A description
★ of the procedure followed is given in Appendix I of the reports of
★ the F.E.S.

In conclusion of this section it must be stressed that we can place no confidence whatsoever on any generalizations made about the parent population from a non-random sample. Randomness is essential when sampling.

3.8 The Distribution of the Sample Mean

Each of the members of our random sample of size n is a *random variable*, where a random variable is the name we apply to all the possible outcomes of the random experiments we introduced at the beginning of this chapter in Section 3.1. Each of these outcomes had a fixed probability of occurring in any trial. Thus for each of the X_i ($i = 1, \ldots, n$) random variables in the sample we could derive the distribution of these probabilities. The mean of the sample, $\bar{X}$, will also be a random variable and so will have its own distribution. This distribution is referred to as the sampling distribution of means, and if we are sampling from a normal distribution with mean μ and variance σ^2 then this distribution of $\bar{X}$ will also be normal with mean μ and variance σ^2/n.

This can be shown as follows:

For any particular sample the sample mean

$$\bar{X} = \frac{1}{n} \sum_{i=1}^{n} X_i$$

where the X_is are independently normally distributed random variables with mean μ and standard deviation σ. The expected value of $\bar{X}$ is

$$E[\bar{X}] = E \left[\frac{1}{n} \sum_{i=1}^{n} X_i \right] = \frac{1}{n} \sum_{i=1}^{n} (E[X_i])$$

$$= \frac{1}{n} \sum_{i=1}^{n} \mu \quad \text{as} \quad E[X_i] = \mu \text{ (for each } i)$$

$$= \mu$$

and the variance of $\bar{X}$ is

$$\text{Var} \left(\frac{1}{n} \sum_i^n X_i \right) = \frac{1}{n^2} \text{Var} \left(\sum_i^n X_i \right) = \frac{1}{n^2} \sum_i^n [\text{Var}(X_i)]$$

$$= \frac{1}{n^2} \sum_i^n \sigma^2$$

$$= \frac{\sigma^2}{n}$$

★ It may be helpful to recall that $\text{Var}(aY) = a^2 \text{Var} Y$, where a is a
★ constant, as
★
★ $$\text{Var}(aY) = \frac{\Sigma(aY - a\bar{Y})^2}{n} = \frac{a^2 \Sigma(Y - \bar{Y})^2}{n} = a^2 \text{Var} Y$$
★

$\sigma/\sqrt{n}$ is often called the *standard error* rather than standard deviation of the mean. Looking at the standard error, we can see that firstly the larger the sample size n, the smaller is the standard error and secondly, the larger the variance σ^2 the larger is the standard error.

Let us take a simple example. The earnings of 10,000 employees of a car manufacturer have a mean of £75 per week and a standard deviation of £25 per week. If we are to take random samples of a hundred employees from the total workforce, the distribution of mean earnings of each of these samples would have mean £75 and standard error, $\sigma/\sqrt{n} = £25/10 = £2.50$. If the samples had each been of twenty-five employees then the distribution of sample means would have had mean £75 and standard error £5.

It is perhaps worth recalling that any statistic which is determined from a random sample has a distribution which we call a sampling distribution; we have concentrated above on the sampling distribution of means.

3.9 The Central Limit Theorem
This is an extremely useful theorem since it allows us to make assertions about the sampling distribution of the sample means

regardless of the form of the distribution of the population. It may be stated as:

If a sample of size n is drawn from a population of mean μ and standard deviation σ then as n approaches infinity the distribution of the sample mean $\overline{X}$ approaches that of a normal distribution with mean μ and standard deviation σ$\sqrt{n}$.

Thus we have the situation where if the original population is not normally distributed, the distribution of sample means corresponds more closely to a normal distribution than does the original population. In practice we often find that the sample size need only be as large as 30 before normality in the sampling distribution becomes apparent. This information will prove very useful because we know the distribution function of the normal distribution (see p. 24) and hence will be able to make probability statements about given differences between the sample mean and the population mean, whether or not the population from which we are sampling is normally distributed.

We now have to look a little more carefully at the properties of the statistics we obtain from random samples and which, we must remember, we use to estimate population parameters.

3.10 Properties of Statistics Obtained from Random Samples —Point Estimators

The point estimator, $\hat{\theta}$ is a function of a set of sample observations, and it is a 'point' in the sense that it refers to a single value and not to a range of values of the variable. For example, the formula for the arithmetic mean of a sample,

$$\overline{X} = \frac{1}{n} \sum_{i=1}^{n} X_i$$

is a point estimator.

★ $\hat{\theta}$, pronounced 'theta hat', is used instead of θ to indicate that we
★   are dealing with an estimator of the population parameter, not the
★   parameter itself.

We must not confuse a point estimator with a point estimate. The latter varies from sample to sample because an estimate is a particular value derived from an estimator. The value of the mean of a particular sample would be an example of an estimate.

When finding an estimate of a population parameter we often have a choice of estimators. The mean, μ, of a population could be estimated using the sample mean, median or mode as an estimator. Faced with such a choice it would seem sensible to choose the estimator best suited to the job. This is not always very easy as estimators can be 'best' in different ways and one has to choose the criterion one wants to decide by. We will mention briefly here four criteria for choosing a best estimator.

(i) Unbiasedness   An estimator $\hat{\theta}$ for a population parameter θ is said to be unbiased if, for all sample sizes, the expected value of $\hat{\theta}$ is equal to the true value of θ. i.e.

$$E[\hat{\theta}] = \theta$$

If $E[\hat{\theta}] \neq \theta$ then $\hat{\theta}$ is said to be biased.

Other things being equal we would prefer an unbiased estimator to a biased one. This may not be the case if say, as is illustrated in Figure 3.1, the unbiased estimator, $\hat{\theta}_2$, were widely dispersed around θ whereas the biased estimator, $\hat{\theta}_1$, had a distribution concentrated near θ.

(ii) Consistency   An estimator $\hat{\theta}$ for a population parameter θ is said to be consistent if the value of the estimator approaches the value of the population parameter as n, the sample size, approaches infinity. For the more mathematically minded this can be written in the sort of way we define a limit, i.e.

For any given number ϵ, where $\epsilon > 0$, the probability $P(|\hat{\theta} - \theta| < \epsilon) \rightarrow 1$ as $n \rightarrow \infty$. ϵ is always in these sorts of definitions taken to be a very small number.

If we take the sample mean as our estimator, then as we increased the sample size from n_1 to n_2 to n_3 we would get sampling

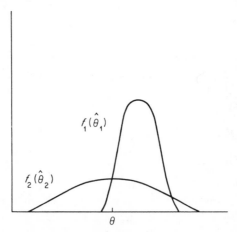

Figure   3.1   Distribution   of   two estimators $\hat{\theta}_1$ and $\hat{\theta}_2$. $f_i(\hat{\theta}_i)$ is the frequency function of $\hat{\theta}_i (i = 1, 2)$

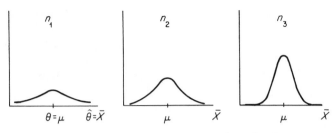

Figure 3.2   Consistency ($n_1 < n_2 < n_3$)

distributions as shown in Figure 3.2. Note the increasing concentration of the sample means around the population parameter μ.

The sample mean is both an unbiased and a consistent estimator, unbiased because $E[\bar{X}] = \mu$ and consistent because as n gets larger the sampling distribution of $\bar{X}$ becomes concentrated around μ. We can see this by considering the standard deviation $\sigma/\sqrt{n}$. As $n \to \infty$, $(\sigma/\sqrt{n}) \to 0$ and hence a concentration around μ.

(iii) Efficiency   An estimator $\hat{\theta}$ for a population parameter θ is said to be efficient if it has the least variance among all possible estimators. If we are just considering two estimators then the more efficient one is that with the smaller variance.

(iv) Sufficiency   An estimator $\hat{\theta}$ for a population parameter θ is said to be sufficient if it provides as much information about θ as any other estimator from the same sample could.

3.11 Properties of Statistics Obtained from Random Samples — Interval Estimators

We have concentrated up till now on estimating a population parameter by a point. This point estimate need not be accurate and so naturally attention focuses on the extent of the possible error that could be associated with our point estimate. What we do is to calculate two numbers, call their difference an interval and then state with a certain amount of confidence that the true population parameter lies within this interval. Two terms are used to describe these intervals — *confidence intervals* or *interval estimates*.

We now look at the way in which we determine these end numbers and we will find that we need to recall some properties of the normal distribution.

If we call our two end numbers A and B, where $B > A$ and both are real, then what we have said above may be expressed as:

$$P(A < \theta < B) = 1 - \alpha \tag{7}$$

the $1 - \alpha$ being a probability associated with the level of confidence

44

required. $1 - \alpha$ is called the confidence coefficient and commonly takes the values 0.99, 0.95 and 0.90. We choose A and B so that we can say that we are for example 95% confident that the true population parameter θ will lie in that interval in the long run.

We continue our investigation by confining our attention to means. We can *standardize* any normal distribution with mean μ and variance σ^2 so that it has a normal distribution with mean 0 and variance 1 by calculating

$$Z = \frac{X - \mu}{\sigma}$$

where X is $N(\mu,\sigma^2)$ and Z is $N(0,1)$. Standardization is thus making the appropriate additive and multiplicative changes to X to obtain our standardized normal variable Z. In the case of $\overline{X}$, where $\overline{X}$ is $N(\mu,\sigma^2/n)$

$$Z = \frac{\overline{X} - \mu}{\sigma/\sqrt{n}}$$

We have shown the sampling distribution of $\overline{X}$ and the standard normal distribution in Figure 3.3. $X - \mu$ is called the *sampling error*. We can easily see from Figure 3.3 that if $z = 1$ then the sampling error is equal to the standard error. The great advantage of using Z is that we have tables of the values of its distribution function widely available. Furthermore the total area under the frequency curve is unity. Hence if we look at Appendix Table A3.2 we can read off the probability that a value drawn from the distribution of Z is greater than some particular value z. This can be seen clearly by looking at Figure 3.4 where we have drawn the distribution of Z. If we take the simplest case where $z = 0$ then from both the top left-hand element in Appendix Table A3.2 and from Figure 3.4 we can see that $P(Z > z) = 0.50$. Similarly from line 11

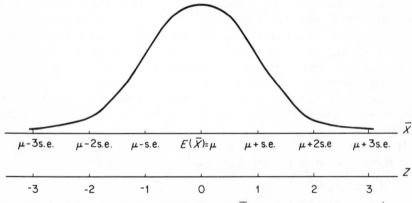

Figure 3.3   A sampling distribution of $\overline{X}$. s.e. is standard error $\sigma/\sqrt{n}$

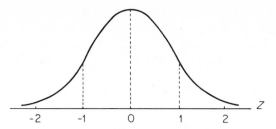

Figure 3.4   The standardized normal distribution

we can see that $P(Z > 1) = 0.1587 = 0.16$ and so on. However, for our confidence intervals we require not just $P(Z > B) = \alpha$ but $P(A < Z < B) = 1 - \alpha$, so we must make more careful use of the tables. Firstly in the case of $P(Z > B) = \alpha$ it is clear that we can fix α, say as 0.1, and hence use the table to find B, 1.28. Since the distribution is symmetric we can easily calculate $P(A < Z < B) = 1 - \alpha$, where $A = -B$. Thus for example if $B = 1, P(-1 < Z < 1) = 0.68$ since we have seen that $P(Z > 1) = 0.16$ and by symmetry $P(Z < -1) = 0.16$. Recalling that our sampling distribution will approach normality as the sample size increases (Central Limit Theorem) we can say that 68% of the items in our sampling distribution will lie between ±1, i.e.

$$P(-1 < Z < 1) = 0.68 \qquad \text{for a standard normal variable}$$

or

$$P\left(-1 < \frac{\overline{X} - \mu}{\sigma\sqrt{n}} < 1\right) = 0.68 \quad \text{for the sampling distribution of } \overline{X}$$

It is easy to think of other examples, if we want our $1 - \alpha$ to be 0.95 then we know that our probability statement for Z will be:

$$P(-1.96 < Z < 1.96) = 0.95$$

and for the sampling distribution of $\overline{X}$

$$P\left(-1.96 < \frac{\overline{X} - \mu}{\sigma\sqrt{n}} < 1.96\right) = 0.95$$

We can rearrange this last statement:

$$P\left(-1.96\frac{\sigma}{\sqrt{n}} < \overline{X} - \mu < 1.96\frac{\sigma}{\sqrt{n}}\right) = 0.95$$

giving

$$P\left(\overline{X} - 1.96\frac{\sigma}{\sqrt{n}} < \mu < \overline{X} + 1.96\frac{\sigma}{\sqrt{n}}\right) = 0.95$$

a probability range for the true population parameter μ.

★ We must make it clear that we cannot state that there is a
★ probability of 0.95 that the true mean lies in the confidence
★ interval. The true mean is a parameter, i.e. of fixed value and so
★ therefore it is either inside the interval or outside it.

We write these intervals in general form as:

$$P\left(\overline{X} - z_c \frac{\sigma}{\sqrt{n}} < \mu < \overline{X} + z_c \frac{\sigma}{\sqrt{n}}\right) = 1 - \alpha \tag{8}$$

and say that the confidence interval or interval estimate for μ is

$$\overline{X} \pm z_c \frac{\sigma}{\sqrt{n}}$$

where z_c is the specified value of Z corresponding to the desired level of confidence, e.g.

$1 - \alpha$	z_c
0.99	2.58
0.95	1.96
0.90	1.645
0.68	1.00
0.955	2.00
0.997	3.00

Our A and B of equation (7) thus become $-z_c(\sigma/\sqrt{n})$ and $z_c(\sigma/\sqrt{n})$ and are known as confidence limits.

Obviously the closer $1 - \alpha$ is to 1 the wider our interval will be, other things being equal, since the size of n affects the interval.

The circumstances under which we can use equation (8) are:

(i) Taking a random sample of size n from a population which is $N(\mu, \sigma^2)$, σ^2 known. The sampling distribution is $N(\mu, \sigma^2/n)$.

(ii) Taking a random sample of size n from a population which is not normal but where σ^2 is known. When n is large we appeal to the Central Limit Theorem, thus overcoming the non-normality of the population problem.

(iii) As soon as σ^2 is unknown we have problems. However, provided we have a large sample ($n > 40$) from a normal population then we can fairly safely use equation (8) if we use

$$s^2 = \frac{\sum\limits_{i}^{n}(X_i - \overline{X})^2}{n - 1}$$

as our estimate of σ^2.

Let us consider why we introduce s^2. The sample variance S^2 may be written as

$$S^2 = \frac{\sum_i^n (X_i - \bar{X})^2}{n}$$

Thus S^2 is a measure of the variability of the sample variables X_i about their mean $\bar{X}$, the sample size being n. S^2 is a function of random variables and hence is itself a random variable.

★ Now we can show that $E[S^2] = [(n-1)/n]\,\sigma^2$, where $\sigma^2 = \sigma^2_{(X)}$

★
★
★
$$E[S^2] = \frac{1}{n} E\left[\sum_i^n (X_i - \bar{X})^2\right] = \frac{1}{n} E\left[\sum_i^n \{X_i - E[X] + E[X] - \bar{X}\}^2\right]$$
★
★
★
$$= \frac{1}{n} \sum_i^n \{E[X_i - E[X]]^2 - E[\bar{X} - E[X]]^2\}$$
★
★
$$= \frac{1}{n} \sum_i^n \left(\sigma^2 - \frac{\sigma^2}{n}\right) \qquad\qquad \text{(N.B. } E[\bar{X}] = E[X])$$
★
★
$$= \frac{1}{n} \left(n\sigma^2 - \frac{n\sigma^2}{n}\right)$$
★
★
$$= \left(\frac{n-1}{n}\right) \sigma^2$$

This tells us that on average the sample variance is smaller than the population variance, σ^2. Since $E[S^2] \neq \sigma^2$ we know that S^2 is a biased estimator of σ^2. We can remove the bias though by multiplying S^2 by an appropriate function of n, i.e.

$$E\left[\frac{nS^2}{n-1}\right] = \sigma^2$$

So

$$\frac{nS^2}{n-1}$$

is an unbiased estimator of σ^2 or

$$\frac{\sum_i^n (X_i - \bar{X})^2}{n-1}$$

is an unbiased estimator of σ^2 and this we write as s^2.

So to recap:

$$S^2 = \frac{\sum\limits_{i}^{n}(X_i - \overline{X})^2}{n} \quad \text{a biased estimator of } \sigma^2.$$

$$s^2 = \frac{\sum\limits_{i}^{n}(X_i - \overline{X})^2}{n-1} \quad \text{an unbiased estimator of } \sigma^2. \tag{9}$$

Obviously we prefer to use an unbiased estimate of σ^2.

We stress that equation (8) holds for large samples from an infinite population or samples with replacement from a finite population.

★　If we draw samples of size n from a finite population size N without
★　replacement then provided we know σ we can use as our interval
★　estimate for μ,
★
★ $$\overline{X} \pm z_c \frac{\sigma}{\sqrt{n}} \sqrt{\frac{N-n}{N-1}}$$
★
★　If σ^2 is unknown then for large n we use $\hat{\sigma}^2$ as an unbiased
★　estimator of σ^2 where $\hat{\sigma}^2 = s^2 (N-n)/(N-1)$.

3.12 Small Samples

We can still use equation (8) if our random sample is small as long as we are sampling from a normal population with known σ^2. When we are sampling from a normal distribution and σ^2 is unknown then we introduce a distribution which is derived from the normal distribution and which is known as the 't' or 'Student's t' distribution (see Section 3.14). Our interval estimator for the population mean μ is then given by:

$$\overline{X} \pm t_c \frac{s}{\sqrt{n}}$$

where t_c is the appropriate value of t and s^2 is the unbiased estimator of σ^2.

In practice the t distribution is an acceptable approximation in cases where the non-normality of the population is not substantial.

3.13 Degrees of Freedom

Before describing the t distribution we need to introduce a new concept, namely degrees of freedom. For our purposes it will suffice to talk about them generally without using too mathematical a definition.

Perhaps if we start by looking at an example the concept will be easier to understand. We take a random sample of n observations, wishing to estimate μ. Now for any given random sample the mean is constant so we can have any values for $n-1$ of the sample measurements but we have no freedom on the last value; it must be such that when added to the other values it equals n times the mean. In this case we say we have $n-1$ degrees of freedom. So in more general terms we can say that in any given sampling situation, the number of degrees of freedom is equal to the number of independent random variables minus the number of constraints on them. Or, in other words, the number of degrees of freedom is given by the number of items in a sample that are active in estimating a particular population characteristic. Equation (9), for example, has $n-1$ degrees of freedom.

3.14 Student's t Distribution
The t distribution is a continuous distribution, symmetric about $t = 0$, its expected value, and is shown in Figure 3.5. The probability associated with a required level of confidence is shown as α as before.

We shall not give the general equation of the probability distribution of Student's t since the areas under the curve are readily available in tabulated form (see Appendix Table A3.3). We shall say though that if we take the sample mean $\overline{X}$ then the values given by

$$t = \frac{\overline{X} - \mu}{s/\sqrt{n}} \tag{10}$$

have a t distribution with $n-1$ degrees of freedom, where $\overline{X}$ and s^2 are the mean and variance of a random sample, size n, from a $N(\mu, \sigma^2)$ population, σ^2 unknown.

As the t ratio given in (10) does not depend upon the population variance σ^2, the t distribution may be used to test hypotheses about the mean μ without knowing the value of the variance σ^2. The main use of the t distribution — and the reason we introduced it — is when small samples are being used.

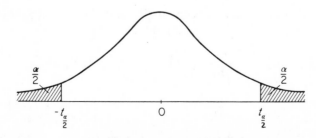

Figure 3.5   The t distribution

3.15 Interval Estimates for Differences of Means

Sometimes we are interested in the differences between two means, say average beef consumption per head in Great Britain and in Japan, or, on a smaller basis, average earned income from a sample of 12 families in Aberdeen and 12 in Exeter.

We would use two different expressions for each of the above examples, but both of the form:

$$\hat{\theta} - z_c \sigma_{\hat{\theta}} < \theta < \hat{\theta} + z_c \sigma_{\hat{\theta}}$$

where as before $\hat{\theta}$ is an estimator of the parameter θ and has an approximately normal sampling distribution. $\sigma_{\hat{\theta}}$ is the standard deviation of $\hat{\theta}$.

For the first case, with two infinite populations we have:

$$\theta = \mu_1 - \mu_2$$
$$\hat{\theta} = \bar{X}_1 - \bar{X}_2$$

and

$$\sigma_{\hat{\theta}} = \sqrt{\left(\frac{\sigma_1^2}{n_1} + \frac{\sigma_2^2}{n_2} \right)}$$

since with $\bar{X}_1$ and $\bar{X}_2$ based on independent random samples of size n_1 and n_2

$$\text{Var}(\hat{\theta}) = \text{Var}(\bar{X}_1 - \bar{X}_2) = \text{Var}(\bar{X}_1) + \text{Var}(\bar{X}_2) = \frac{\sigma_1^2}{n_1} + \frac{\sigma_2^2}{n_2}$$

where σ_1^2 and σ_2^2 are the population variances. This gives the interval estimator as:

$$\bar{X}_1 - \bar{X}_2 - z_c \sqrt{\left(\frac{\sigma_1^2}{n_1} + \frac{\sigma_2^2}{n_2} \right)} < \mu_1 - \mu_2 < \bar{X}_1 - \bar{X}_2$$
$$+ z_c \sqrt{\left(\frac{\sigma_1^2}{n_1} + \frac{\sigma_2^2}{n_2} \right)}$$

an extension of equation (8).

With small samples and unknown population variances, the second case, we would use the t distribution and so would have:

$$\bar{X}_1 - \bar{X}_2 - t_c s_c \sqrt{\left(\frac{1}{n_1} + \frac{1}{n_2} \right)} < \mu_1 - \mu_2 < \bar{X}_1 - \bar{X}_2$$
$$+ z t_c s_c \sqrt{\left(\frac{1}{n_1} + \frac{1}{n_2} \right)}$$

where t_c is the appropriate value of t with $(n_1 + n_2 - 2)$ degrees of freedom and s_c^2 is the combined variance of the samples, i.e.

$$s_c^2 = \frac{(n_1 - 1)s_1^2 + (n_2 - 1)s_2^2}{n_1 + n_2 - 2}$$

References and Suggested Reading
1. An introduction to set theory can be found in many elementary books on statistics. The interested reader could for example consult

R. E. Walpole, *Introduction to Statistics*, 2nd Edition, Collier--Macmillan, London, 1974, Ch. 2.

2. In this chapter we introduced a major source of statistical information about the expenditure patterns of households in the U.K.

Family Expenditure Survey, H.M.S.O., London, annually.

We also mentioned the *Business Monitor Series*, which is an important source of data about U.K. industries. It is also published by H.M.S.O., but the intervals between publication vary with the different industries.

Questions
1   It is found that 60% of people between the ages of 18 and 21 years pass their driving test on the first attempt. At the second attempt the pass rate is 75% of the first attempt pass rate. At the third attempt the pass rate is only half that of the second attempt. What is the probability of anyone in that age range failing their test (a) the first, (b) the second, (c) the third time?

2   Suppose we take a random sample of size 9 from a normal population with known variance $\sigma^2 = 4$. The mean μ is not known and we wish to calculate an interval estimate for the mean.

Our sample is: 12, 8, 11, 13, 10, 12, 11, 14, 13

How should we proceed?

3   A firm manufacturing light bulbs claims each bulb has an average life of 1,200 hours. 10 are selected at random from the firm's production unit and tested. The sample mean was found to be 1,130 hours and the sample standard deviation 100 hours. Does the 95% confidence interval for the population mean μ cover the claimed mean of 1,200 hours?

4   An economist is conducting a survey on the costs involved in using disposable paper sheets in hospitals. As part of his enquiry he finds, after taking a large number of observations, that in a certain ward the average life of a paper sheet is 60 hours with a standard deviation of

25 hours. The distribution of hours lasted seems to be approximately normal.

One sheet lasts 85 hours, what is the probability that it will last 100 hours or more?

5   An ironmonger's shop includes in its annual sale 21 factory-packed saucepans. It is known that 9 of them are red in colour, 8 are blue and 4 are yellow although the colour tags have come off. A customer buys 3 at a bargain price. What is the probability that all 3 saucepans will be the same colour?

6   An egg importer finds that 15% of the cartons of 6 eggs are damaged. A carton is picked at random, checked and returned to the consignment. This procedure is repeated a further 3 times. What is the probability that out of the 4 cartons so inspected 3 were undamaged and 1 was damaged?

7   (a) In how many ways can a shopper choose one or more brands of jam from a batch containing 5 different brands?
(b) Let us alter this to types of jam and suppose that our shopper is faced with choosing 7 different jams from 10
    (i) How many choices are there?
    (ii) If we know 4 will be picked, how many ways can the 7 be selected?
    (iii) If we know that at least 4 of a given 5 will be picked, how many ways can our shopper select 7?

Answers

1(a) Probability(pass first time) = 0.60 $P(P_1)$

therefore P(fail first time) = 0.40 $P(F_1)$

(b) The probability of an individual failing the second time is conditional on their having failed the first time, i.e. $P(F_2 | F_1)$

Now $P(P_2 | F_1) = 0.60(0.75) = 0.45$

therefore $P(F_2 | F_1) = 0.55$

Probability someone fails two tests = $P(F_1)P(F_2 | F_1)$

$$= 0.40(0.55)$$

$$= 0.22$$

(c)   To fail the third test an individual has to have failed the second test, i.e. $P(F_3 | F_2)$ is needed.

Now $P(P_3 | F_2) = 0.45(0.5) = 0.225$

therefore $P(F_3 | F_2) = 0.775$

Probability someone fails three tests = $P(F_1)P(F_2 \mid F_1)P(F_3 \mid F_2)$

$$= 0.40(0.55)(0.775)$$

$$= 0.1705$$

2   Sample mean $\bar{X} = 11.56$
Our interval estimate is

$$11.56 - z_c \frac{2}{\sqrt{9}} < \mu < 11.56 + z_c \frac{2}{\sqrt{9}}$$

with a confidence level of 95% we have

$$11.56 - 1.96 \left(\frac{2}{3}\right) < \mu < 11.56 + 1.96 \left(\frac{2}{3}\right)$$

$$10.25 < \mu < 12.87$$

with a confidence level of 99% we have

$$11.56 - 2.58 \left(\frac{2}{3}\right) < \mu < 11.56 + 2.58 \left(\frac{2}{3}\right)$$

$$9.84 < \mu < 13.28$$

showing that an increase in the confidence level leads to a wider interval.

We cannot say whether the interval estimates given contain the true mean; what we can say is that if we were to take a 100 samples we would expect (in the first case) 95 of the resulting interval estimates to cover the true mean.

N.B. We could also have used this method had our sample been large $(n \geqslant 30)$ and the sampling been carried out without replacement from an infinite population or with replacement from a finite population. In either case we would have assumed $s = \sigma$ as explained in the text.

3   We have both a small sample and an unknown population variance and so we use the t distribution.

The relevant value of t for $10 - 1 = 9$ d.f. and 95% confidence level is 2.26.

The confidence interval is given by:

$$1{,}130 - 2.26 \frac{(100)}{(\sqrt{10})} < \mu < 1{,}130 + 2.26 \frac{(100)}{(\sqrt{10})}$$

$$1{,}130 - (0.7147)(100) < \mu < 1{,}130 + 71.47$$

$$1{,}058.53 < \mu < 1{,}201.47$$

which covers the claimed mean of 1,200 hours — just.

If the sample size is increased to 15 with $\overline{X} = 1,130$ and $s = 100$, then the new t value would be 2.14 and the 95% confidence interval:

$$1,130 - 2.14\,\frac{(100)}{(\sqrt{15})} < \mu < 1,130 + 2.14\,\frac{(100)}{(\sqrt{15})}$$

$$1,130 - 55.25 < \mu < 1,130 + 55.25$$

$$1.074.75 < \mu < 1,185.25$$

which does not cover the claimed mean.

4   This problem involves conditional probability and the standard normal distribution.

Let our random variable for the number of days the sheet lasts be X. We assume that X is normally distributed with $\mu = 60$ and $\sigma = 25$.

Now the probability that the sheet lasts 100 hours or more is conditional on it having lasted more than 85 hours, therefore

$$P(X \geqslant 100) = P(X \geqslant 100 \text{ and } X \geqslant 85)$$
$$(\text{i.e. } P(X \geqslant 100) = P(X \geqslant 100 \cap X \geqslant 85))$$

So we can write

$$P(X \geqslant 100 \mid X \geqslant 85) = \frac{P(X \geqslant 100)}{P(X \geqslant 85)}$$

We can obtain $P(X \geqslant 100)$ and $P(X \geqslant 85)$ from the standard normal distribution. So changing from the x scale to the z scale we get:

$$x = 100,\ z = \frac{100 - 60}{25} = \frac{8}{5} = 1.6, \quad x = 85,\ z = \frac{85 - 60}{25} = 1.0$$

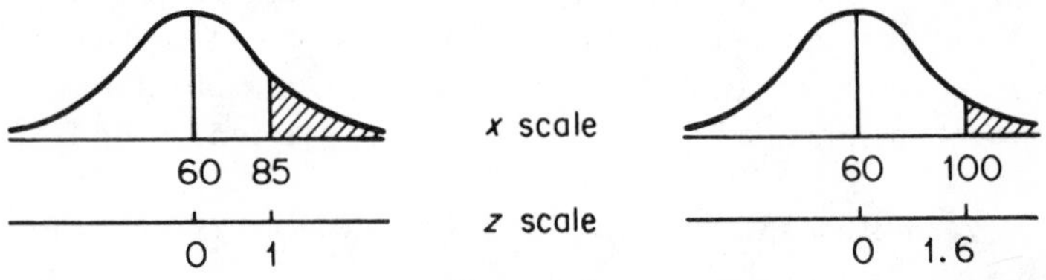

Figure 3.6

$$P(X \leqslant 100) = P(Z \geqslant 1.6) = 0.5000 - 0.4452 = 0.0548$$

$$P(X \leqslant 85) = P(Z \geqslant 1.0) = 0.5000 - 0.3413 = 0.1587$$

Therefore

$$P(X \geqslant 100 \mid X \geqslant 85) = \frac{0.0548}{0.1587} = 0.3453$$

i.e. if a sheet lasts 85 hours there is a probability of 0.35 that it will last 100 hours or more.

5   This problem involves combinations and probability.
The total number of ways of selecting 3 saucepans from 21 = $_{21}C_3$
The number of ways of selecting 3 red = $_9C_3$, etc. Therefore

$$P(\text{all 3 red}) = \frac{_9C_3}{_{21}C_3}$$

$$P(\text{all 3 blue}) = \frac{_8C_3}{_{21}C_3}$$

$$P(\text{all 3 yellow}) = \frac{_4C_3}{_{21}C_3}$$

$P(3 \text{ saucepans are the same colour})$

$$= \frac{1}{_{21}C_3}(_9C_3 + {_8C_3} + {_4C_3})$$

$$= \frac{6}{21(20)(19)}\left[\frac{9(8)(7)}{6} + \frac{8(7)(6)}{6} + 4\right]$$

$$= 0.1083$$

6   We had 4 draws from the consignment of cartons so we could have had 0, 1, 2, 3 or 4 damaged cartons. In other words the number of damaged cartons is a random variable. In fact we have 1 damaged carton but we do not know whether this occurred on the first, second, third or fourth draw. Since 15% of the whole consignment is damaged we know that the probability of drawing an undamaged one is $1 - 0.15 = 0.85$

Hence probability damaged carton occurs on

First draw = 0.15 x 0.85 x 0.85 x 0.85 = 0.0921
Second draw = 0.85 x 0.15 x 0.85 x 0.85 = 0.0921
Third draw = 0.85 x 0.85 x 0.15 x 0.85 = 0.0921
Fourth draw = 0.85 x 0.85 x 0.85 x 0.15 = 0.0921

since all draws are independent.

$$P(1 \text{ damaged}) = 0.0921 + 0.0921 + 0.0921 + 0.0921$$
$$= 0.3684$$

(Another way of considering why we have to consider 4 occasions is quite simply as follows: we have that the total number of outcomes = 2^4 .. ... 2 mutually exclusive events — damaged or undamaged — and 4 of these outcomes lead to the required result.)

7   (a) There are two ways of looking at this problem.

(i) Saying that you want one or more implies that the shopper can choose 1 or 2 or 3 or 4 or 5 different brands. There are $_5C_1$ ways of choosing 1, $_5C_2$ ways of choosing 2, etc; and so the number of ways of choosing 1 or more brands

$$= {_5}C_1 + {_5}C_2 + {_5}C_3 + {_5}C_4 + {_5}C_5$$
$$= 5 + 10 + 10 + 5 + 1$$
$$= 31$$

(ii) The other, and maybe easier way is to say that with each brand the shopper can do one of two things — take it or leave it. The total number of outcomes is hence 2^5, but this includes the possibility of rejecting all 5 brands and we want 1 at least to be chosen.

Hence required number of ways = $2^5 - 1$
$$= 31$$

(b)   (i) $_{10}C_7 = \dfrac{10(9)(8)}{3(2)} = 120$

(ii) $_6C_3 = \dfrac{6(5)(4)}{3(2)} = 20$

(iii) If all 5 are picked then we have $_5C_2 = 10$ ways.
If just 4 are picked then choose these in $_5C_4 = 5$ ways and remaining 3 in $_5C_3 = 10$ ways, i.e. $5(10) = 50$ ways. Total number of ways = $10 + 50 = 60$.

4 Hypothesis Testing

4.1 Types of Hypotheses

We now turn to another branch of statistical inference, the theory of hypothesis testing. We must remember that we are dealing with a *statistical* hypothesis, i.e. a statement about the frequency distribution of a random variable. Thus a remark of the form 'Butter will be in short supply next year' is a hypothesis while the remark 'Most people think that butter will be in short supply next year' is a statistical hypothesis, the difference being that the latter is a hypothesis concerning the behaviour of an observable random variable, namely the percentage of people involved (most). We should also want to define our terms 'most' and 'short supply' precisely.

Sometimes we talk about a parametric hypothesis. This is on occasions when we have specified the underlying distribution of the observations and the hypothesis relates purely to the value(s) of its parameter(s). If, however, we have a hypothesis where no parameter is specified then we say we have a non-parametric hypothesis. For example, the hypothesis that the average expenditure of patrons in a restaurant is £5 is a parametric hypothesis whereas the hypothesis that the distribution of their expenditure on weekdays is identical to the distribution at weekends is a non-parametric hypothesis.

We can be even more specific and talk about simple and composite hypotheses. The distinction here concerns the number of parameters of the distribution which have their values specified by the hypothesis. If a distribution has m parameters, the hypothesis specifies values for l of these: then if $l = m$ the hypothesis is simple, whilst if $l < m$ then the hypothesis is composite. In the previous chapter we introduced degrees of freedom and it is interesting to note here that the number of degrees of freedom of the hypothesis is given by $m - l$. l is the number of constraints.

'Testing' a hypothesis is, simply, a method of deciding whether to accept or reject the hypothesis in question.

4.2 Specification of Hypotheses

Before we can test our statistical hypothesis we have to ensure that it is carefully specified. We cannot test the hypothesis that the average gross earnings of female manual workers is low, but we can test the hypothesis that the average gross earnings of female manual workers is £22.60 per week (the mean value of the sample taken in the 1973 *New Earnings Survey*), from a random sample of such workers. We must, too, specify our hypothesis and then decide whether to accept or reject it on the basis of our data. We must not let our data specify our hypotheses for us.

4.3 The Null Hypothesis

We are dealing with the basic situation of using a random sample to provide information about the parameters of the parent population. If, for example, our sample statistic is the sample mean, then we know from the last chapter that the sample means have a probability distribution of their own and that we can make probabilistic statements about various values of $\overline{X}$. We can say, for example, that there is a 1 in 20 chance of a value lying in a particular range in that sampling distribution of means. An example here will help to clarify the issue. We have chosen one that we all know something about so that we have a 'feel' about the result.

Suppose we wish to test the hypothesis that a particular coin is a fair one. We tend to call the hypothesis that we wish to test the *null hypothesis*. With a fair coin we know that we would expect in the long run to find that the probability, p, of a head, say, was a half. Let us denote the null hypothesis by H_0, so

H_0 : P(a head) = ½

Let the number of tosses be $n = 100$ and the outcome be 38 heads. If we assume the coin is fair then we can use the normal approximation to the binomial distribution to find the probability of getting 38 or less heads in a series of 100 throws.

★ The binomial distribution is a discrete probability distribution
★ which describes many real-life events and hence is frequently used.
★ A binomial situation occurs when we have a fixed number of
★ statistically independent trials n and a constant probability of 'a
★ success' on each trial, say p. The only other outcome on each trial is
★ 'a failure' and has probability $1 - p$ of occurring. The terms
★ 'success' and 'failure' are commonly used but merely mean that an
★ event has or has not taken place; we could equally well use 1 and 0
★ or 'yes' and 'no' depending on the particular situation.

★ The probability of x successes in n trials is given by
★
★ $$f(x) = {_nC_x}\, p^x\, (1-p)^{n-x} \quad x = 0, 1, 2, \ldots, n$$
★
★ The mean of the binomial distribution, $\mu = np$ and the variance
★ $\sigma^2 = np(1-p) = npq$, where $q = 1-p$. We use the fact that when n
★ is large we can use the normal distribution to approximate binomial
★ probabilities (R. E. Walpole, *Introduction to Statistics* gives a full
★ explanation of the binomial distribution).

So

$n = 100$

$\mu = np = (100)(\tfrac{1}{2}) = 50$

$\sigma = \sqrt{(npq)} = \sqrt{[(50)(\tfrac{1}{2})]} = 5 \quad$ where $q = 1-p$

Therefore z our standard normal variable is

$$\frac{38.5 - 50}{5} = -\frac{11.5}{5} = -2.3$$

Thus, using Appendix Table A3.2, we can see that the probability of getting 38 or less = 0.0107.

★ Since we are approximating a discrete distribution to a continuous
★ one we must use 38.5 rather than 38 because the probability that
★ the value 38 is taken is equal to the area under the normal curve
★ between 37.5 and above and less than 38.5. (For a fuller
★ explanation see R. E Walpole, *Introduction to Statistics*.)

This whole example is shown in Figure 4.1.

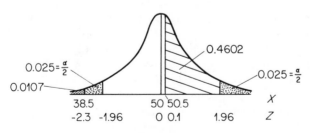

Figure 4.1

0.0107 is undoubtedly a very low probability and it tells us that either an exceedingly rare event has occurred or that we were wrong to assume that the coin was fair. What we are concerned with is the decision process in the latter case. If the outcome had been 51 heads then our z value would have been $(50.5 - 50)/5 = 0.1$ and the probability of getting 51 or more heads would have been 0.4602. We would be quite happy to assume that the coin was fair on this result. What we must decide to do is to determine in advance the decisive level of probability for each test of a hypothesis. This decisive level is usually denoted by α. This is the same α as that introduced in Chapter 3. As with confidence intervals we can 'centre' our $1 - \alpha$ leaving $\alpha/2$ in each 'tail' of the distribution, and this is shown in the figure. If before starting to test this hypothesis we had decided that our α would be 0.05 and hence $\alpha/2 = 0.0250$, then we would have rejected H_0, the null hypothesis, on the basis of the first outcome because $0.0107 < \alpha/2$, and accepted it on the basis of the second outcome because $0.4602 > \alpha/2$. Because we accept the null hypothesis if our computed value of Z lies between ± 1.96 this area under the normal curve is often called the region of acceptance or non-significance. The complementary region, in this case the two tails, is called the region of rejection or significance. The use of the word 'significance', which by the way gives rise to the common label for α of *'significance level'*, arises in a rather loose way from the idea of only accepting a hypothesis because nothing can be found against it, whereas a hypothesis is rejected because some reason can be found for rejecting it. The whole situation is sometimes likened to that of a prisoner who is assumed innocent until proved guilty.

It is worth noting at this point that the term 'significant' is commonly used to indicate significance at the 0.05 or 5% level and 'highly significant' to indicate significance at the 0.01 or 1% level.

4.4 The Alternative Hypothesis

In the previous example of tossing a coin we said our null hypothesis H_0 was $P(\text{a head}) = \frac{1}{2}$. Now if we had felt unable to accept H_0 (as with the first outcome of 38 heads) then we would have accepted the alternative hypothesis that $P(\text{a head}) \neq \frac{1}{2}$. This we will denote by H_1. The concept of null and alternative hypotheses can be difficult to understand. Perhaps the best approach is to think of the null hypothesis as a 'maintained' or 'working' hypothesis — one which will be held to be true until and unless sufficient evidence is obtained to reject it in favour of some alternative. For example, if we are interested in whether a new advertising campaign to attract depositors to building societies is more successful than the previous one, then we assume there is no difference between the two and test that null hypothesis.

We have looked at one form that the alternative hypothesis, H_1, can take, but there are others. Instead of having $P(\text{a head}) \neq \frac{1}{2}$, we could have considered $p > \frac{1}{2}$, $p < \frac{1}{2}$ or $p = 0.55$ for example, although the alternative hypothesis of $p \neq \frac{1}{2}$ allows for the possibility that $p > \frac{1}{2}$ or $p < \frac{1}{2}$. If we consider a general population parameter θ, taking a specific value θ_0 under the null hypothesis, then we can list different null hypotheses and their corresponding alternative hypotheses as:

Null hypothesis		Alternative hypothesis	
$H_0:\ \theta = \theta_0$	(1a)	$H_1:\ \theta \neq \theta_0$	(1b)
$H_0:\ \theta \leqslant \theta_0$	(2a)	$H_1:\ \theta > \theta_0$	(2b)
$H_0:\ \theta \geqslant \theta_0$	(3a)	$H_1:\ \theta < \theta_0$	(3b)

In the terminology explained in Section 4.1 we can say that hypothesis (1a) is a simple hypothesis, while hypothesis (1b) represents a composite hypothesis.

4.5 Examples of Hypotheses

(i) A simple economic relationship is given by $Q = a + bP$, a linear demand function, where Q is the quantity of the good demanded, P is the price of the good and a and b are constants. We would expect this function to take the form shown in Figure 4.2. We would thus expect to find if we were estimating a linear demand function, that $b < 0$. We would test $H_0:\ b = 0$ against the alternative hypothesis $H_1:\ b < 0$.

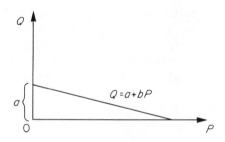

Figure 4.2 A simple linear hypothesis

(ii) Suppose we consider the same sort of very simple relationship as we have in (i) above, but choosing as our problem this time the effect on income for an individual of an extra year's education. We would then have a simple relationship of the form $Y = a + bE$, where Y is the income of an individual who has E years of education, a and b are constants, and b is the income from an extra year's education. Here we would expect to find that $b > 0$ and we would test $H_0:\ b = 0$ against the alternative hypothesis, $H_1:\ b > 0$.

(iii) A company has to decide whether or not to install a different computer. Obviously there would be no point in changing over if the costs of such a move were greater than the benefits. The hypothesis which we make our null hypothesis will depend on the particular situation. If it is the case that the changeover costs are low, that the new computer will have similar running costs to the existing one and that the existing one has been a bit unreliable, then our hypotheses would be

H_0: The new computer is as good as or better than the existing one.
H_1: The present computer is better than the new computer would be.

If, on the other hand, the present computer has been reliable, its running costs are slightly lower than the different model and changeover costs are going to be high — a new building may be needed for example — then we would alter our hypotheses putting the burden of proof on the new model to show that it really would be an improvement. Hence our hypotheses would be

H_0: The present computer is as good as or better than the new one.
H_1: The new computer would be an improvement over the present one.

(iv) A drug company is proposing to market a new drug. We might expect the procedure to be that we test the null hypothesis that the drug has no effect against the alternative hypothesis that the drug is effective. However, because we could be dealing with something that potentially could be very harmful, e.g. thalidomide, it would be more sensible to assume the new drug is harmful until it is proved otherwise. The null hypothesis would then be that the drug was harmful and the alternative hypothesis would be that the drug was acceptable.

4.6 One-Tailed and Two-Tailed Tests

We have seen that H_1 can take various forms. When testing our null hypothesis H_0 we will find that we use a 'one-tailed' test or a 'two-tailed' test depending on the form of H_1. We have already mentioned 'two tails' in our example with the coin but we can now be more explicit. We remind ourselves that hypothesis testing is a decision-making process — we take a sample from a population, calculate a sample statistic, prescribe a region which we have previously called 'significant' or region of rejection but which is commonly called the 'critical' region, say R, which is such that H_0 is rejected if our sample statistic falls inside R and accepted otherwise. Thus, by prescribing a region we are prescribing a test of a hypothesis. In other words our critical region consists of sample points, which, if observed, would lead us to reject H_0. In our example with the coin we in fact had as our critical region sample values of 40 and below and 60 and above since these are the regions defined by the relation P(sample statistic is a member of $R|H_0$) = 0.05 which was the level of α we suggested. Here we have an example of a two-tailed test which we can state generally as

H_0: $\theta = \theta_0$
H_1: $\theta \neq \theta_0$

i.e. our alternative hypothesis is that θ may be either less than θ_0 or greater than θ_0.

If for example we were testing means then we would have

$$H_0 : \mu = \mu_0$$
$$H_1 : \mu \neq \mu_0$$

where μ_0 is a given value of the population mean μ. We would reject H_0 if our sample mean $\bar{X}$ was either 'too large' or 'too small' compared with μ_0. If we are going to keep α as our level of significance and hence $1 - \alpha$ as our area of non-significance then we have to divide our α equally between the two tails — our critical region. This is shown in Figure 4.3.

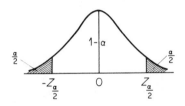

Figure 4.3   Critical region for a two-tailed test

The computation involved would be for

$$Z = \frac{\bar{X} - \mu_0}{\sigma / \sqrt{n}}$$

where n is the size of the random sample, σ^2 is the known population variance and $\bar{X}$ was normally distributed. If the test were at the 5% level of significance then H_0 would be rejected if the calculation yielded a value of Z which was such that $Z > 1.96$ or $Z < -1.96$ (using Appendix Table A3.2 we can see that $P(Z > 1.96) = 0.025 = \alpha/2$).

A one-tailed test arises in situations where our critical region is concentrated entirely in either the upper or the lower tail, e.g.

$$H_0 : \theta = \theta_0 \qquad H_0 : \theta = \theta_0$$
$$\text{or}$$
$$H_1 : \theta > \theta_0 \qquad H_1 : \theta < \theta_0$$

respectively.

Following our previous example with means we would have for a lower-tailed test,

$$H_0 : \mu = \mu_0$$
$$H_1 : \mu < \mu_0$$

This time we would reject H_0 if the sample mean $\bar{X}$ was too low compared with μ_0 (see Figure 4.4). Testing at the 5% level of

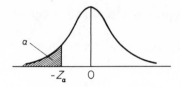

Figure 4.4   Critical region for
a one-tailed test

significance, H_0 would be rejected if our calculated value of Z fell below -1.645.

For an upper-tailed test at 5%, H_0 would be rejected if $Z > 1.645$.

The critical region R is — in both figures — the shaded area. The most commonly used levels of significance give rise to the following critical values for the test statistic Z.

α	Two-tailed test	One-tailed test
10%	1.645	1.28
5%	1.96	1.645
1%	1.58	2.33

4.7 Differences Between Means

A very common problem involves testing to see whether or not there is a significant difference between the means of two given distributions. Examples are:

(1) Is there a significant difference between the average number of persons engaged in General Stores in the North and in the West Midlands?

(2) Is there a significant difference between the average amount of public investment in new construction of hospitals per head of population in the South East and the South West?

(3) Is there a significant difference between the sales of gas per head in Wales and in Scotland?

(4) Is there a significant difference between the percentage of income spent on cheese by families with gross incomes of £65 per week where (a) there are children under 10 years and (b) there are children of 10 years and above?

(5) Is there a significant difference between the average earnings of non-manual women aged 18 and over in (1) the Rural West of the West Midlands region and (2) Fylde of the North West region? Let us actually work through this example using data from Table 72 of the *Abstract of Regional Statistics*, 1972. The data on average earnings are distinguished for males and females, manual and non-manual workers; the numbers in the sample are given as are the standard errors of the estimated averages. Using the two areas we have to set up a suitable null

hypothesis, and we would use a two-tailed test if we were willing to accept that there might be a difference between the two population means but not to specify the direction of the difference on the basis of our knowledge of the two areas. The first area contains places such as Shrewsbury, Ludlow and Clun and the second, Blackpool and Lytham St Annes. They are clearly different in character but we would have to know a great deal about both earnings of non-manual women aged 18 and over and their distribution between various occupations in these sorts of areas before we could say that we thought that the difference between the two means, μ_1 and μ_2 respectively, would be positive rather than negative or vice-versa. Thus we set up the null hypothesis $H_0: \mu_1 - \mu_2 = 0$ and test this hypothesis of no difference between the means against the alternative hypothesis $H_1: \mu_1 \neq \mu_2$.

Rural West		Fylde
$\bar{X}_1$ = £19.7		$\bar{X}_2$ = £18.2
n_1 = 114		n_2 = 110
$s_1 / \sqrt{n_1}$ = £0.8		$s_2 / \sqrt{n_2}$ = £0.6
μ_1	Population mean	μ_2

Our test statistic is

$$Z = \frac{(\bar{X}_1 - \bar{X}_2) - (\mu_1 - \mu_2)}{\sqrt{(\sigma_1^2 / n_1 + \sigma_2^2 / n_2)}}$$

where $\bar{X}_1$, $\bar{X}_2$ are the sample means, n_1, n_2 the sizes of the two samples, s_1, s_2 the sample standard deviations and σ_1^2, σ_2^2 the known population variances (the variance of the difference between two means was explained in Section 3.14). However, where the samples are large, s^2 and σ^2 can be equated. Using large samples, of course, ensures that the distribution of the sample mean is approximately normal.

Hence in this case

$$z = \frac{19.7 - 18.2}{\sqrt{(0.64 + 0.36)}} = 1.5$$

At the 5% level of significance the critical values of Z for a two-tailed test are ± 1.96. Our value of Z lies inside this range and hence we accept the null hypothesis at the 5% level of significance that there is no difference between the average earnings of non-manual women in Rural West, West Midlands and Fylde, North West.

We can write our hypotheses in the general form:

$$H_0: \mu_1 - \mu_2 = 0$$

and

$$H_1: \underbrace{\mu_1 \neq \mu_2}_{\text{two-tailed test}} \quad \text{or} \quad \underbrace{\mu_1 > \mu_2 \quad \text{or} \quad \mu_1 < \mu_2}_{\text{one-tailed test}}$$

4.8 Small Samples

If our sample is small and is drawn from a population which is approximately normally distributed, then we have to change our test statistic to t from Z. When dealing with a single mean, say where H_0: $\mu = \mu_0$, then our test statistic would be

$$t = \frac{\bar{X} - \mu_0}{s/\sqrt{n}}$$

where s is the known sample standard deviation and n is the size of the small sample (n usually $\leqslant 30$). The appropriate value of t is found from the t tables (Appendix Table A3.3) at significance level α and $n-1$ degrees of freedom. For a two-tailed test H_0 will be rejected if the calculated value of t is such that $t < -t_{\alpha/2}$ or $t > t_{\alpha/2}$. For an upper-tailed test then H_0 would be rejected if $t > t_\alpha$ and similarly if $t < -t_\alpha$ for a lower-tailed test.

For completeness we give the t test to be used when testing for a significant difference between means with small samples n_1 and n_2. However, since the samples have to be drawn from independent normal distributions with equal standard deviations it cannot often be used. To test H_0: $\mu_1 = \mu_2$ we use

$$t = \frac{\bar{X}_1 - \bar{X}_2}{\sqrt{[(n_1 - 1)s_1^2 + (n_2 - 1)s_2^2]}} \sqrt{\left[\frac{n_1 n_2 (n_1 + n_2 - 2)}{n_1 + n_2}\right]}$$

The appropriate number of degrees of freedom here is $n_1 + n_2 - 2$.

4.9 Proportions

The procedure for testing hypotheses which involve proportions deserves our attention since they too are often used. Consider a unit trust group which is trying to decide whether to float a new trust. It carries out a survey itself and finds that 30% of those interviewed say they would participate in such a trust if it were formed. This percentage is considered rather high and so the group commissions a survey firm to determine the views of a random sample of people. The null hypothesis H_0: $p = 0.3$ is tested against the alternative H_1: $p < 0.3$ where p is the proportion of those who will purchase.

The results of the random sample survey of 300 people were that 79 of them said they would invest in such a trust. If we accept H_0 then the expected number of 'buyers' out of 300 will be $np = 300 (0.3) = 90$ and the variance will be $npq = 300 (0.3)(0.7) = 63$ since we have a binomial situation approximated to the normal distribution on account of the sample size.

Our test statistic Z can in general be written as:

$$Z = \frac{X - np}{\sqrt{(npq)}}$$

and in particular

$$z = \frac{79 - 90}{7.937} = -1.3859$$

Thus H_0 would be rejected at the 10% level of significance but accepted at the 5% level of significance.

4.10 Differences Between Proportions

The theory for testing the null hypothesis of no significant difference between two population proportions, i.e. $H_0 : p_1 - p_2 = 0$ against the alternative hypothesis, say $H_1 : p_1 - p_2 \neq 0$, is quite straightforward. We take two large independent samples

	Sample 1	Sample 2
Number in sample with a particular attribute	X_1	X_2
Size of sample	n_1	n_2

The two population proportions are p_1 and p_2 and if we take the difference between the two sample proportions $[(X_1 / n_1) - (X_2 / n_2)]$ then its sampling distribution will be approximately normal with mean $p_1 - p_2$ and variance $[(p_1(1-p_1)/n_1) + (p_2(1-p_2)/n_2)]$. Since we do not usually know the values of p_1 and p_2 we obtain their estimator, $\hat{p}$, by combining the data from the two samples, i.e.

$$\hat{p} = \frac{X_1 + X_2}{n_1 + n_2}$$

This makes the estimator of the variance of the sampling distribution

$$= \frac{\hat{p}(1-\hat{p})}{n_1} + \frac{\hat{p}(1-\hat{p})}{n_2}$$

and so

$$\hat{\sigma}_{p_1 - p_2} = \sqrt{\left[\hat{p}(1-\hat{p}) \left(\frac{1}{n_1} + \frac{1}{n_2} \right) \right]}$$

Our test statistic

$$Z = \left(\frac{X_1}{n_1} - \frac{X_2}{n_2} \right) \Big/ \sqrt{\left[\hat{p}(1-\hat{p}) \left(\frac{1}{n_1} + \frac{1}{n_2} \right) \right]}$$

4.11 Type I and Type II Errors

At the beginning of this chapter, in our coin example, we calculated the probability of getting 38 heads or less. Our answer was 0.0107 and we concluded that either a very rare event had occurred or that we were wrong to assume that the coin we tossed was a fair one. Later we defined α as our significance level and used this as a mark against which to compare our test statistic. If we do commit the error of rejecting the null hypothesis when it is in fact true (i.e. a very rare event) then we say we have committed a Type I error. Hence the probability of committing Type I error $= \alpha$. The other possible error, that of accepting a hypothesis when it is false, is called a Type II error. We call the probability of a Type II error β. The decision making process can be summarized as:

	Actual situation	
Decision	H_0 true	H_0 false
Accept H_0	Correct	Type II error
Reject H_0	Type I error	Correct

Because these are errors the decision maker will want to ensure that the probabilities of making these errors are as low as possible. It could be costly if they were otherwise. However, we have the problem that the two types of errors are interdependent, e.g. for a given sample size reducing the probability of a Type I error has the effect of increasing the probability of a Type II error. The following example will illustrate this point.

The salaries of a large number of female employees in an organization are known to be normally distributed with mean $\mu_0 = £1,248$ and standard deviation $\sigma = £180$. A new system of work assessment and payment is introduced and a sample taken of 36 female employees under the new system gives an average salary $\bar{X} = £1,302$. We will assume the standard deviation is unaltered. If we were testing to see if the difference between $\bar{X}$ and μ_0 were significant then we would test

$$H_0 : \mu = \mu_0$$

against

$$H_1 : \mu > \mu_0$$

where μ is the population mean for the new system. Our Type I error, α, would be stated and would commonly be 0.05 or 0.01. In order to calculate our Type II error we have to have a specific alternative hypothesis, in other words we have to say exactly how the null hypothesis is false. μ could be equal to £1,275, £1,298, £1,318 or indeed any figure within reason and the Type II error will alter with the change in μ.

Suppose we pick the last of these values and say that the new system does give higher salaries and that the mean for all female employees under the new system is £1,318. Testing

$$H_0 : \mu = \mu_0$$

against

$$H_1 : \mu > \mu_0$$

leads us to reject H_0 at the 5% level of significance and accept it at the 1% level since

$$z = \frac{1,302 - 1,248}{180/\sqrt{36}} = \frac{54}{30} = 1.80$$

The critical value at the 5% level, C_5, is given by

$$\frac{C_5 - 1,248}{180/\sqrt{36}} = 1.645$$

$$C_5 = 1,248 + 49.35 = 1297.35$$

The critical value at the 1% level, C_1, is given by

$$\frac{C_1 - 1,248}{180/\sqrt{36}} = 2.33$$

$$C_1 = 1,248 + 69.9 = 1,317.9$$

If we consider the 5% level first, i.e. the probability of a Type I error is 0.05, then if our observed value of $\bar{X}$ exceeds 1,297.35, H_0 will be correctly rejected. We will commit a Type II error if the observed value of $\bar{X}$ is less than 1,297.35, leading us to accept H_0 when it is in fact false. So taking $H_1 : \mu = £1,318$ as the true alternative hypothesis we can calculate our Type II error:

$$P(\bar{X} < 1,297.35) = P\left(Z < \frac{1,297.35 - 1,318}{180/\sqrt{36}} \right)$$

$$= P(Z < 0.6883)$$

$$= 0.2451$$

i.e. Type II error at $\alpha = 0.05$ is 24.51%. This situation is shown in Figure 4.5.

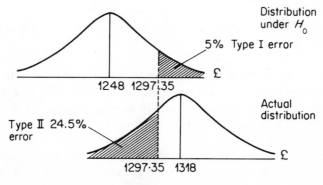

Figure 4.5

Now consider the 1% level of significance. We commit a Type II error this time if the observed value of $\bar{X}$ is less than 1,317.9.

$$P(\bar{X} < 1{,}317.9) = P\left(Z < \frac{1{,}317.9 - 1{,}318}{180/\sqrt{36}}\right)$$

$$= P(Z < 0.0033)$$

$$= 0.5000$$

So Type II error with $\alpha = 0.01$ is 50%. This situation is shown in Figure 4.6 where the two distributions have been superimposed on each other.

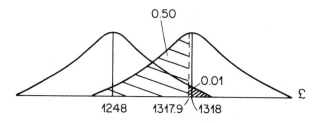

Figure 4.6

Summarizing our results emphasizes the relative movements:

Type I	Type II
0.05	0.2451
0.01	0.5000

So given our sample size we can reduce the Type I error by changing the point at which we decide whether to accept or reject H_0, but this increases the Type II error.

It is possible to reduce both errors simultaneously by increasing the sample size but this presumably would not be a costless exercise.

4.12 Power
The power of a test is defined to be equal to $1 - P$ (Type II error), i.e.

Power = $1 - \beta$

This is the probability of rejecting H_0 when it is false. Obviously the greater the power the smaller the Type II error. We can also define power using previous notation, i.e. P(sample statistic is a member of $R \mid H_1) = 1 - \beta$, and this makes it obvious that power is a function of H_1. We can draw what is known as a power curve which is in fact a graph of the power function of a region R, as is shown in Figure 4.7.

Figure 4.7 is often described as a power function of a test; it was pointed out earlier in this chapter that prescribing a region was

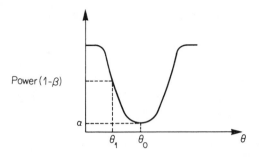

Figure 4.7

equivalent to prescribing a test of a hypothesis. What we have shown in Figure 4.7 is a general power function testing H_0: $\theta = \theta_0$ against H_1: $\theta \neq \theta_0$. The power curve shows the power of the test for all possible values of θ and the given value of α. We can read off the probability of rejecting H_0 when H_1: $\theta = \theta_1$ is true (i.e. the power of the test/region with respect to H_1: $\theta = \theta_1$) on the vertical scale. When θ takes the value θ_0 the power of the test is equal to α, the probability of a Type I error.

The power function helps us to choose our critical region R so as to give us a 'good' test. Obviously we want a test whose error probabilities, α and β, are as small as possible. We have seen that we cannot choose R in such a way that α and β are simultaneously minimized, but we can compromise by controlling the probability of a Type I error by restricting ourselves to a small fixed value of α, and then choose the region which has the greatest power, hence minimizing the probability of a Type II error, β. If we can choose a region which has greatest power for all possible alternative hypotheses on parameter θ, then the test of H_0 using that region is called uniformly most powerful (U.M.P.).

The ideas expressed in the previous paragraph are due to Neyman and Pearson and are more fully discussed in statistics texts (see suggested reading).

References and Suggested Reading
1. We have introduced two further sources of statistical information on the U.K.

New Earnings Survey, Department of Employment, H.M.S.O., annually.
Abstract of Regional Statistics, H.M.S.O., annually.

The first of these is a detailed sample survey of earnings of employees in most industries and occupations. The second is a collection of statistics with a regional breakdown. Many of these data are available elsewhere.

2. For a further examination of the binomial distribution and the normal approximation to it see, for example,

R. E. Walpole, *Introduction to Statistics*, 2nd Edition, Collier–Macmillan, London, 1974.

3. The discussion of power tends to be omitted from elementary books. The reader might, however, try

A. M. Mood and F. A. Graybill, *Introduction to the Theory of Statistics*, McGraw–Hill, New York, 1963

or

S. D. Silvey, *Statistical Inference*, Penguin Books, Harmondsworth, 1970, pp. 96–97.

However, much of this book is relatively mathematical.

Questions

1. A small shopkeeper feels that he may be losing trade to a supermarket and decides to issue trading stamps as an incentive. He calculates that average sales per customer over the month before trading stamps are given are £2.48 with a standard deviation of £1.61. In the month after the introduction of trading stamps the average sales per customer from a simple random sample of 100 customers was £2.75. The standard deviation was unchanged. Has the shopkeeper benefited from the introduction of trading stamps?

2. Is there a 'significant' difference between the average amount a private household in the United Kingdom spends on 'furniture, including repairs' (F) and 'gas and electric appliances, including repairs' (GEA)? Use the data given below which comes from Appendix 2 of the 1971 *Family Expenditure Survey*.

	Average weekly household expenditure	Percentage standard error	Number of households recording expenditure
Furniture, including repairs	0.37	10.8	751
Gas and electric appliances, including repairs	0.45	6.5	2175

3. Nine manual workers (men) were asked to state their gross earnings for one week in October 1973. These were:

£44.06 41.27 42.96 37.52 64.33 57.81 40.90 35.36 38.77

Test the hypothesis at the 5% significance level that these manual workers were employed in manufacturing industries where the average earnings for a week in October 1973 were £41.52 for men. (Table 162, *Monthly Digest of Statistics*, June 1974.)

4. The 'acceptable' level of unemployment is always a controversial subject. An opinion research firm decides to carry out a poll in such a way as to determine whether there is any significant difference between the proportion of people (a) over 25 years and (b) 25 years and under who consider 3.4% of the labour force to be an 'acceptable' level of unemployment.

The results of independent random samples of the two age groups are:

	Over 25 years	25 years and under
Number who thought 3.4% 'acceptable'	90	115
Sample size	225	225

Test the hypothesis $H_0: p_1 = p_2$ against the alternative $H_1: p_1 \neq p_2$ where p_1 and p_2 are the population proportions for 'over 25 years' and '25 years and under' respectively. Comment on your result.

5. We think that the average weekly expenditure on food by households with total weekly incomes in the '£60 and under £70' group is £15. However, as we do not know the income and price elasticities of demand of this group for food, the actual value could be higher or lower than this. If we decided to take a random sample of households in this income group to test our belief,

(a) what size should our sample have been, and
(b) what should the decision rule have been

if we are prepared to accept a Type I error of 0.05 and a Type II error of 0.10, but in fact $\sigma^2 = £42$ and the actual average weekly expenditure on food for this group is £14.20?

Answers
1. We treat the data from the time period before the issue of stamps as our population data. We set up our null hypothesis as:

$$H_0: \mu \leqslant 2.48$$

i.e. the shopkeeper has not benefited. Our alternative hypothesis is:

$$H_1: \mu > 2.48$$

We test to see whether it is probable that the random sample was drawn from the pre-stamp population. We decide to test at the 5%

significance level, i.e. we are prepared to take the risk of being wrong 5 times in every 100. No particular significance level was given and so the conventional level of 5% was chosen.

Now

Sample size, $n = 100$

Sample mean, $\overline{X} = 2.75$

Sample standard deviation, $\dfrac{\sigma}{\sqrt{n}} = \dfrac{1.61}{10} = 0.161$

z, our standard normal variable, $= \dfrac{2.75 - 2.48}{0.161}$

$$= \frac{0.27}{0.161}$$

$$= 1.677$$

The critical value for Z for a one-tailed test at the 5% significance level is 1.645. Our calculated value of Z is greater than this and so we reject H_0 at this level.

Hard-and-fast conclusions are not easy to draw. The type of shop is not specified but given that it is compared with a supermarket it seems likely to be selling food. The time periods involved are two consecutive months — relatively short, but how much of the rise in average sales is due to an increase in prices? Will the effect last or is the increase due mainly to novelty value? Before we can say whether or not the shopkeeper has benefited we would have to know the cost to him of issuing trading stamps.

Finally it is worth pointing out that although our result was 'significant', it was not 'highly significant', had we chosen 1% as our level we could not have rejected H_0.

2. We are testing to see whether there is a 'significant' difference between two population means say μ_F and μ_{GEA} for the two commodities. The two samples are certainly large enough for us to equate the sample variance, s^2, with the population variance σ^2 and to ensure that the distributions of the sample means are approximately normal.

Our null hypothesis H_0: $\mu_F = \mu_{GEA}$ is tested against the alternative H_1: $\mu_F \neq \mu_{GEA}$. The two-tailed test is used because there are no *a priori* reasons for expecting either commodity to have more spent on it on average than the other.

Our test statistic Z is given by:

$$\frac{\overline{X}_F - \overline{X}_{GEA}}{\sqrt{(s_F^2/n_F + s_{GEA}^2/n_{GEA})}}$$

It is important to note that the F.E.S. gives 'percentage standard errors' which means that to find the actual standard error for each commodity we must multiply the percentage standard error by the average expenditure and divide by 100. Therefore

$$z = \frac{0.37 - 0.45}{\sqrt{\{[(10.8)(0.37)/100]^2 + [(6.5)(0.45)/100]^2\}}}$$

$$= \frac{-0.08}{\frac{1}{100}\sqrt{(15.9680 + 8.5556)}}$$

$$= \frac{-0.08}{\frac{1}{100}\sqrt{(24.5236)}} = -\frac{0.08}{0.04952}$$

$$= -1.6155$$

The critical values for Z for a two-tailed test at the 5% level are ± 1.96. Our value of Z falls inside this range and we therefore accept the null hypothesis that there is no significant difference between the amount private households spend on average on furniture and on gas and electric appliances.

3. The size of the sample is small, 9, and so we have to use t as our test statistic. Provided we assume that the population from which the sample was drawn is approximately normally distributed then we can test

$$H_0: \mu = 41.52$$

against

$$H_1: \mu \neq 41.52$$

using a test statistic given by

$$t = \frac{\overline{X} - 41.52}{s/\sqrt{n}}$$

where $\overline{X}$, s and n are the sample mean, standard deviation and size respectively.

From the sample $\overline{X} = 44.78$

$$s = \sqrt{\frac{n}{(n-1)}} \sqrt{\left(\frac{\Sigma X_i^2}{n} - \overline{X}^2\right)} = \sqrt{\frac{9}{8}} \sqrt{\left(\frac{18,804.406}{9} - 2,005.284\right)}$$

$$= 9.73$$

and $\sqrt{n} = 3$, therefore our calculated value of t

$$= \frac{44.78 - 41.52}{9.73/3}$$

$$= 1.01$$

The critical value for t corresponding to a two-tailed test at the 5% significance level is 2.31, which is greater than our calculated value and so we accept the null hypothesis. These nine men were likely to have been employed in manufacturing industries.

A point to note is that we assumed that the population of earnings was normally distributed. This is not in fact strictly true but it is safe to assume that this would not materially affect the result.

4.   We have first to decide what Type I error to accept. An $\alpha = 0.05$ would seem reasonable in this situation giving us critical values of Z at ± 1.96 given our test is two-tailed.

Our estimator $\hat{p}$ of p_1 and p_2 is given by

$$\hat{p} = \frac{90 + 115}{225 + 225} = \frac{205}{450} = \frac{41}{90}$$

The estimator of the standard deviation, $\hat{\sigma}_{p_1 - p_2}$, of the sampling distribution is given by

$$\hat{\sigma}_{p_1 - p_2} = \sqrt{\left[\frac{41}{90} \left(1 - \frac{41}{90} \right) \left(\frac{2}{225} \right) \right]}$$

$$= \frac{7}{(90)(15)} \sqrt{82}$$

Our test statistic

$$z = \left(\frac{90}{225} - \frac{115}{225} \right) \Big/ \frac{7\sqrt{82}}{(90)(15)}$$

$$= -\frac{25}{225} \Big/ \frac{7\sqrt{82}}{(90)(15)} = \frac{-(10)(15)}{7\sqrt{82}}$$

$$= -2.37$$

At the 5% significance level we therefore reject H_0 and accept the hypothesis that there is a significant difference between the proportion of people in the two age groups concerning their attitude towards levels of unemployment.

The figure 3.4% was taken, being twice the rate suggested by the Manchester Inflation Workshop as the 'natural' level of unemployment. It is equivalent (in 1973–4) to about 800,000 people. It is an interesting comment that the results would almost certainly be different if the figure of 800,000 rather than 3.4% had been used.

5.   We have the situation as shown in Figure 4.8. The critical values of Z for $\alpha = 0.05$ (2 tails) are ± 1.96 and for $\beta = 0.10$ the critical value is

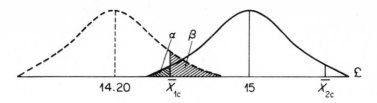

Figure 4.8

1.28. Thus

$$-1.96 = \frac{\overline{X}_{1c} - 15}{\sqrt{(42/n)}}$$

where n is the required sample size and

$$1.28 = \frac{\overline{X}_{1c} - 14.20}{\sqrt{(42/n)}}$$

therefore

$$\overline{X}_{1c} = 15 - 1.96\sqrt{\frac{42}{n}} = 14.20 + 1.28\sqrt{\frac{42}{n}}$$

$$\sqrt{\frac{42}{n}}(1.28 + 1.96) = 15 - 14.20 = 0.80$$

$$n = \left(\frac{3.24}{0.80}\right)^2 42$$

$$= 688.91$$

A sample of 689 households in this income group would therefore give us the ability to test the hypothesis that the population mean was equal to £15 at the 0.05 significance level (i.e. $\alpha = 0.05$) and with a probability of 10% of accepting the null hypothesis when it was in fact false (i.e. $\beta = 0.10$). The power of the test $1 - \beta = 0.9$, i.e.. there is a chance of 90% of rejecting the null hypothesis when it was in fact false.

(b) The decision rule for a null hypothesis of £15 average weekly expenditure would be: 'Accept $H_0: \mu = 15$ if sample mean lies between $\overline{X}_{1c}$ and $\overline{X}_{2c}$', i.e. if sample mean is less than £15.48 and greater than £14.52.

5 The Relationship Between Two Variables

Up to this point we have only considered statistics relating to one variable at a time. We now have sufficient resources at our command to look at much more interesting economic problems concerning the interaction of two variables.

5.1 Graphical Examination

Our economic theory tells us that we would expect to find a relationship between imports and national income. We shall write that imports are a function of national income:

$$M = f(Y) \tag{1}$$

where M stands for imports and Y for national income. Clearly this is a very crude aggregate relationship because some imports are finished products and hence substitutes for domestic products whereas others are materials which cannot be found in the importing country. We therefore expect that there will be a tendency for imports to rise with income, but that the relationship will not be perfect.

If we plot the values of the two variables on a graph, often referred to as a *scatter diagram*, with imports on the vertical axis and income on the horizontal we can immediately obtain an idea of the accuracy of our assertion. In Figure 5.1 we have done this for the U.K. using quarterly data over the period 1968—1970. The variables are defined as follows:

M — Imports — imports of goods and services in £'000 mn
Y — Income — Gross Domestic Product at factor cost in £'000 mn

Our response to the situation shown in Figure 5.1 is that it would not appear unreasonable to suggest that imports and income do tend to vary together. It would however be useful to obtain some measure of

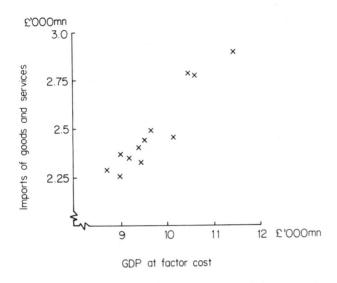

£'000mn

Imports of goods and services

GDP at factor cost

Figure 5.1   Scatter diagram of imports and GDP of the U.K., 1968 to 1970. Note the choice of axes to show a reasonable spread to the data. Source: *Economic Trends*, **216**, Oct. 1971, H.M.S.O.

the degree to which the two vary together, since merely looking at a diagram is rather *ad hoc*.

5.2 The Covariance

If we wanted to look at the variation in the two variables separately during the sample period we could look at their variance or standard deviation. An equivalent measure in two dimensions is the covariance. In any one time period we have a pair of observations, one on imports and one on income. We can phrase this differently by saying that we have one observation from the bivariate distribution of imports and income. The joint mean of the sample we have drawn from this distribution is $(\overline{Y}, \overline{M})$.

★ For those who have forgotten their school mathematics: we can
★ refer to any point on a two-dimensional graph by reading off its
★ coordinates. We firstly trace a vertical line from the point to the
★ horizontal axis and read off the value, say y, and then trace a
★ horizontal line across to the vertical axis and read off the value, say
★ m. We can then refer to the point as having a value (y, m).

We can consider where each of the individual points (Y_t, M_t) lie relative to the mean and measure the distances $(M_t - \overline{M})$ and $(Y_t - \overline{Y})$. In Table 5.1 we have set out the values of M and Y used in constructing Figure 5.1 and have computed in columns (3) and (4) the values of

Table 5.1
Imports and income, U.K., 1968—70, quarterly, £'000 mn

Time t	(1) Imports M_t	(2) Income Y_t	(3) $M_t - \bar{M}$	(4) $Y_t - \bar{Y}$	(5) $(M_t - \bar{M})^2$	(6) $(Y_t - \bar{Y})^2$	(7) $(M_t - \bar{M})(Y_t - \bar{Y})$
1968 Q1	2.28	8.75	−0.198	−1.044	0.0392	1.0899	0.2067
Q2	2.25	9.00	−0.228	−0.794	0.0520	0.6304	0.1810
Q3	2.35	9.22	−0.128	−0.574	0.0164	0.3295	0.0735
Q4	2.29	9.66	−0.188	−0.134	0.0353	0.0180	0.0252
1969 Q1	2.37	9.09	−0.108	−0.704	0.0117	0.4956	0.0760
Q2	2.41	9.54	−0.068	−0.254	0.0046	0.0645	0.0173
Q3	2.44	9.66	−0.038	−0.134	0.0014	0.0180	0.0051
Q4	2.44	10.30	−0.038	0.506	0.0014	0.2560	−0.0192
1970 Q1	2.49	9.69	0.012	−0.104	0.0001	0.0108	−0.0012
Q2	2.78	10.43	0.302	0.636	0.0912	0.4045	0.1921
Q3	2.77	10.66	0.292	0.866	0.0853	0.7500	0.2529
Q4	2.87	11.53	0.392	1.736	0.1537	3.0137	0.6805
Σ	29.74	117.53	0.004[a]	0.002[a]	0.4923	7.0809	1.6898
Σ/N	2.478	9.794	0.000	0.000	0.0410	0.5901	0.1408

[a]Note size of rounding error.

$(M_t - \overline{M})$ and $(Y_t - \overline{Y})$. Columns (5) and (6) proceed in the standard manner to calculate the squared deviations round the mean, from which the respective variances have been calculated in the last line of the table. In column (7), however, we have considered the relationship between the deviation from the mean of each variable with the deviation of the other and computed $(M_t - \overline{M})(Y_t - \overline{Y})$. Whereas the squared deviation round the mean is necessarily non-negative, the product of the deviations has no such constraint. If the result is positive then both variables are either greater than their respective means or both are less than their means. If the result is negative then one is greater and the other less. This is shown in Figure 5.2, where the origin of the graph is $(\overline{Y}, \overline{M})$. Positive values of the product result from observations in quadrants B and C and negative values from observations in A and D.

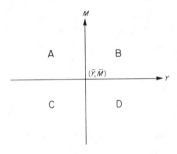

Figure 5.2

However, this tells us only about each individual observation. If we calculate the mean of these products,

$$\frac{\displaystyle\sum_{t=1}^{N} (M_t - \overline{M})(Y_t - \overline{Y})}{N}$$

we can see what the average product is. If this average is positive then the observations tend to lie in quadrants B and C and if it is negative they tend to lie mainly in A and D. Finally, if the average is zero then the observations are spread over all four quadrants. This average is called the *Covariance* of M and Y and is written Covariance(M, Y) or Cov(M, Y) for short. All three possibilities are shown in Figure 5.3.

5.3 Correlation
There are a number of drawbacks about using the covariance as a measure of the relationship between variables. Firstly, it is measured in the product of the units of the two component variables and hence is not comparable with other distributions using other variables. If

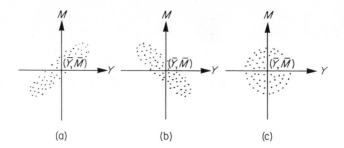

Figure 5.3 (a) Positive covariance. (b) Negative covariance. (c) Zero covariance

imports were measured in £mn instead of £'000 mn then clearly Cov(M, Y) would be 1,000 times as large. Secondly, we have no idea of how closely the two variables are related, any more than we had by looking at the original scatter diagram in Figure 5.1. A large covariance may merely reflect the smallness of the units. We have no means of telling whether there is any important difference between a positive (or negative for that matter) covariance and zero. We require therefore a statistic which

(a) tells us how closely the two variables are associated,
(b) is unit free and comparable, and
(c) tells us whether the association is positive or negative.

The statistic which is used for this purpose is the correlation coefficient. However, before we explain its nature we must note a major restriction in its application. In (1) we made no assumption about the nature of the relationship between Y and M. In considering the covariance we have assumed implicitly that the relationship falls within certain limits. Y and M could be exactly related in such a way that all the observations lie on the circumference of a circle, as in Figure 5.4, and yet their covariance would be zero. We have assumed

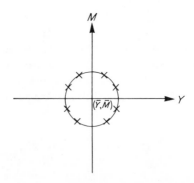

Figure 5.4

implicitly that the functional relationship can be represented by a straight line, or something fairly closely approximating to it. In using the correlation coefficient we are seeing how closely the relationship conforms to a straight line. It is by no means impossible to see how closely the relationship conforms to various other curves or even step functions, but at this stage we will confine ourselves to examining the closeness of the linear relationship between variables.

★ It is possible to represent any straight line by an equation of the
★ form $Y = a + bX$ where Y is measured along the vertical axis and X
★ along the horizontal axis. The parameter a is the value at which the
★ line cuts the vertical axis and b is the slope of the line, i.e. the
★ amount that Y changes when X increases by one unit.

It is easy to think of a case where M and Y have an exact linear relationship, for example

$$M = Y \qquad (2)$$

In this case the covariance is equal to each of the variances

$$\text{Cov}(M, Y) = \text{Var}(M) = \text{Var}(Y) \qquad (3)$$

We require a statistic where this is the limiting case of positive relationship. Similarly if

$$M = -Y \qquad (4)$$

then

$$\text{Cov}(M, Y) = -\text{Var}(M) = -\text{Var}(Y) \qquad (5)$$

and this is the limiting case of perfect negative relationship.
The *correlation coefficient*, r, has the following form

$$r = \frac{\text{Cov}(M, Y)}{\sqrt{[\text{Var}(M) \, \text{Var}(Y)]}} \qquad (6)$$

If (2) holds then $r = 1$ and if (4) holds then $r = -1$. Thus $-1 \leqslant r \leqslant 1$. Secondly if there is no relationship $\text{Cov}(M, Y) = 0$ and hence $r = 0$.
At first glance (6) may seem unnecessarily complicated, since either $\text{Cov}(M, Y)/\text{Var}(M)$ or $\text{Cov}(M, Y)/\text{Var}(Y)$ would have the same properties. However this relies upon the fact that M and Y are measured in the same units. Also (2) is only one out of an infinite number of possible exact linear relationships between M and Y whose general form is

$$M = a + bY \qquad (7)$$

If we ignore a for the time being and consider

$$M = bY \qquad (8)$$

we can substitute into (6) for M as follows

$$r = \frac{\text{Cov}(bY, Y)}{\sqrt{[\text{Var}(bY)\,\text{Var}(Y)]}} \tag{9}$$

$$= \frac{b\text{Cov}(Y, Y)}{\sqrt{[b^2\,\text{Var}(Y)\,\text{Var}(Y)]}} \tag{10}$$

$$= \frac{b\text{Var}(Y)}{b\text{Var}(Y)} = 1 \tag{11}$$

Clearly $r = -1$ if $M = -bY$.

The actual value of a is immaterial to the analysis because we are dealing in terms of deviation round the mean. We have shown earlier (p. 15) that $\text{Var}(X) = \text{Var}(X + 20)$. It, therefore, would not matter to our conclusions if there had been a systematic under-recording of imports during the entire period of some £200,000 per quarter. It is only variable errors that affect the correlation coefficient, and unfortunately it is only too likely that it is these that we have in practice.

Returning to our example, we can see from the last line of Table 5.1 that $\text{Var}(M) = 0.0410(\text{Col}(5))$, $\text{Var}(Y) = 0.5901(\text{Col}(6))$ and $\text{Cov}(M, Y) = 0.1408(\text{Col}(7)) -$ all in £'000 mn^2. Hence we can compute r

$$r = \frac{0.1408}{\sqrt{[(0.0410)(0.5901)]}} = \frac{0.1408}{0.1555} = 0.905$$

It is worth emphasizing at this stage that we can make an elementary check on the accuracy of our calculations: i.e. $-1 \leqslant 0.905 \leqslant 1$. Our initial scatter diagram also led us to expect $r > 0$, as does economic theory, and our expectations are thus borne out.

5.4 Testing the Correlation Coefficient

Just because we have a positive value for r in our example we cannot immediately say that there is definitely a positive relationship between Y and M. r is a statistic calculated from a sample and hence we cannot say anything about the population correlation coefficient, ρ, until we know something about the sampling distribution of r. $E[r] = \rho$ for all values of ρ, thus r is an unbiased estimator, but its variance depends upon the value of ρ. Let us begin by taking the common case where $\rho = 0$. Here the distribution of r is symmetric, in fact

$$\frac{(r - \rho)}{\sqrt{[(1 - r^2)/(n - 2)]}}$$

is distributed as t with $(n - 2)$ degrees of freedom. Thus the appropriate test statistic, where H_0 is $\rho = 0$ and H_1 is $\rho \neq 0$, is

$$t = \frac{r - \rho}{\sqrt{[(1 - r^2)/(n - 2)]}} \tag{12}$$

In our example

$$t = \frac{0.905}{\sqrt{(0.181/10)}} = \frac{0.905}{0.135} = 6.73$$

If the size of the test we require is 5% then we can look up $t_{0.05}$ for 10, $(n - 2)$, degrees of freedom in Appendix Table A3.3, and establish that since

$$t(10)_{0.05} = 2.23 < 6.73 = t$$

we reject H_0. This confirms our original hypothesis that the relationship between income and imports is approximately linear.

It will be immediately obvious to the reader that if $\rho \neq 0$ the sampling distribution of r will no longer be symmetric, since r can only hold values in the range $-1 \leqslant r \leqslant 1$. Hence while the distribution has the form shown in Figure 5.5(a) if $\rho = 0$ when $\rho > 0$ it becomes negatively skewed as in Figure 5.5(b). Fortunately it is possible to transform r to z by the following relation:

$$z = \frac{1}{2}\log_e\frac{(1 + r)}{(1 - r)} \tag{13}$$

where z is approximately normally distributed with mean

$$\zeta = \frac{1}{2}\log_e\frac{(1 + \rho)}{(1 - \rho)} \tag{14}$$

and variance

$$\frac{1}{n - 3}$$

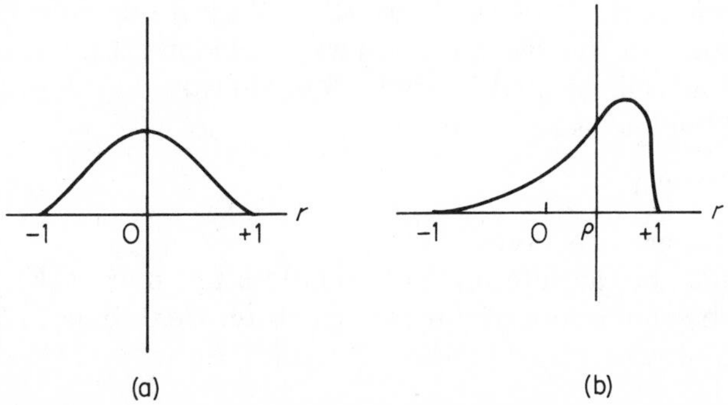

(a)　　　　　　　　　　(b)

Figure 5.5　Sampling distribution of r. (a) $\rho = 0$. (b) $\rho > 0$

The derivation of these results is rather complex and has therefore been omitted (the original exposition is given in Fisher (1915)). The test statistic is now

$$z^* = \frac{z - \zeta}{\sqrt{[1/(n-3)]}} \qquad (15)$$

where z^* is $N(0, 1)$.

It is perhaps also worth noting that we can test for the difference between two sample correlation coefficients, r_1 and r_2 in a way analagous to testing for the difference between two means or proportions (see pp. 64 − 67). The difference $z_1 - z_2$ is approximately normally distributed with mean zero and variance

$$\frac{1}{(n_1 - 3)} + \frac{1}{(n_2 - 3)}$$

under the null hypothesis that $\rho_1 = \rho_2 = \rho$ and hence $\zeta_1 = \zeta_2 = \zeta$, where n_1 and n_2 are the two sample sizes.

5.5 Stochastic Relationships

Up to this point we have considered relationships between variables without any real concern for their form. We have merely calculated a correlation coefficient to see if there is a linear association between two variables. However, we can distinguish between two general sorts of relationship, which we shall call *deterministic* and *stochastic*. A deterministic relationship is one where the equality between the two sides of the equation necessarily holds. For example, if a country has a unified income tax structure where the first £k of income are free of tax and all income above that is taxed at the rate t, total income and disposable income will be perfectly correlated

$$r_{Y,YD} = 1$$

where Y stands for total income and YD for disposable income, for all people earning more than £k. If we drew a scatter diagram of a sample of observations on the two variables we would have a set of points on a straight line as is illustrated in Figure 5.6. We would therefore write the equation of the line as

$$YD = b + wY \qquad (16)$$

where b and w are constants.

This would be a deterministic relationship since (16) is exactly determined by the nature of our tax structure. We define

$$YD \equiv Y - T \qquad (17)$$

where T is the tax paid, and our tax structure can be written as

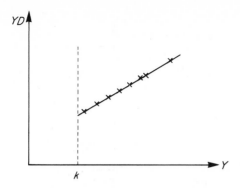

Figure 5.6 Scatter diagram between income and disposable income

$$T = t(Y - k) \tag{18}$$

Substituting for T in (17) from (18) we obtain

$$YD = Y - t(Y - k)$$
$$= Y - tY + tK$$
$$= tK + (1 - t)Y \tag{19}$$

Since both t and k are fixed let us set $tk = b$ and $(1 - t) = w$ and we can see that by substituting these two values in (19) we obtain (16). Hence the relationship $YD = f(Y)$ is deterministic.

If at the same time we calculated the correlation between the consumption, C, of the people in the sample and their disposable income we would obtain a value in the range

$$0 \leqslant r_{C,YD} \leqslant 1$$

The scatter diagram of the observations is shown in Figure 5.7(a). From this sample we would accept the hypothesis of a linear association between C and YD, but the points on the scatter diagram do not all lie on the line

$$C = a + cYD \tag{20}$$

yet (20) is most certainly the form the economist adopts for a simple consumption function. We have to say in this case that the relationship between any particular value of disposable income, YD_i, and the corresponding value of consumption, C_i, has the form

$$C_i = a + cYD_i + e_i \tag{21}$$

where e_i is the difference between the value of C calculated from (20) and its actual value

$$e_i = C_i - (a + cYD_i) \tag{22}$$

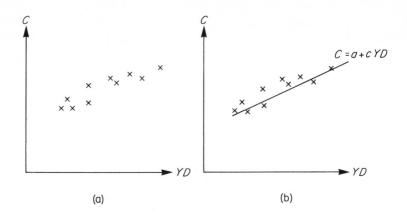

Figure 5.7   Scatter diagram between consumption and disposable income

The relationship between disposable income and consumption is *stochastic* in that (20) is the basic relationship that we believe holds (from the Greek 'στοχος' meaning 'aim' or 'conjecture'). Thus while we may observe

$$C = a + cYD + e \tag{23}$$

we think that

$$E[C \mid YD] = a + cYD \tag{24}$$

and hence

$$E[e] = 0 \tag{25}$$

In economics a stochastic relationship is a description of human behaviour. We do not believe that people behave exactly alike, and observation supports this, but we do believe that we can generalize about people's actions. The residual term, e, represents all the many other factors that affect people's decisions. In expressing (23) we think that overall these extra factors will offset each other and in total have an expected value of zero. We could therefore draw a straight line through the points in Figure 5.7(a) as is shown in Figure 5.7(b) and this line represents the basic relationship (20). In the next chapter we shall discuss how we might decide exactly where to draw this line.

The particular assumptions we make about the residuals are very important. We are assuming that no coherent factor has been left out. However, in (20) household size affects how households spend their incomes. All other factors being constant, the more people there are in a household the greater the proportion of their income they would tend to spend on consumption. In this case we know something about e and therefore may feel that $E[e] \neq 0$.

It is easy to think of examples of deterministic relationships where a residual exists; but in such cases the residual represents a variable that has been omitted from the equation. In our taxation example, (16), for instance, if we had made no allowance for transfer payments we would find that although there was a positive correlation between income and disposable income it was not perfect. This does not mean that (16) now becomes a stochastic relationship, merely that it is untrue, and the correct equation should be

$$YD = b + wY + TR \tag{26}$$

where TR stands for transfer payments.

We shall develop this problem further in the next three chapters, but at present we wish to establish that:

a deterministic relationship has the form

$$Y = f(X) \tag{27}$$

and a stochastic relationship has the form

$$Y = f(X) + e \tag{28}$$

where

$$E[Y \mid X] = f(X) \tag{29}$$

SIMPLE TIME SERIES ANALYSIS

5.6 Economic Time Series
In this section we shall be concerned with the performance of economic variables over time. We can divide up the sorts of samples we obtain of economic phenomena into three categories:

(i) Those which are observed over a period of time and have values for each time interval.
(ii) Different events which are measured at the same time.
(iii) A combination of (i) and (ii).

These three categories are referred to as:

(i) Time series data.
(ii) Cross-section data.
(iii) Pooled data.

In the previous section we considered the general stochastic relationship of the form

$$Y = a + bX + e \tag{30}$$

We now have the situation where we have obtained a set of observations on a variable over time. Thus we can rewrite (30) as

$$Y = f(t) + e \qquad (31)$$

where t refers to time, to obtain a general expression of the behaviour of a variable over time. The economist often wishes to obtain a measure of the movement of an economic time series, and much of this wish is related to the rate of change of a variable. We wish to answer questions like: How fast have prices risen during the past year? How fast did wage rates rise during the 1960s? However, we can also just consider change over time rather than rate of change; often when we graph an economic series over time the residual element is sufficiently strong that it is difficult to decide what the underlying movement in the series is. This underlying movement is called a 'trend'. We can approach the problem of determining a trend from two points of view; we can either decide on a particular form of trend and apply it to the data or we can average out the fluctuations in the series and see what the shape of the resultant trend is like.

5.7 Imposing a Particular Form of Trend
5.7.1 A Linear Trend
Let us take the case of imposing a particular form of a trend on the data first. The simplest hypothesis is a linear trend

$$Y = a + bt + e \qquad (32)$$

In Figure 5.8 we show a graph of national income over the period 1950—71. The effects of inflation have been removed from the series by expressing each year in terms of 1963 prices. (A description of the method of deflation from current prices to constant prices is given on p. 205.) The next step is to attempt to draw a straight line through the data. The easiest method is to use a ruler and draw a line by eye; however, everybody will draw a fractionally different line, so the points must lie near the line for this method to be at all useful. Clearly we ought to adopt a method of calculation whereby anyone looking at these data will draw precisely the same line. We shall describe such a method — ordinary least-squares regression — in the next chapter. At this point it is worth noting the *ad hoc* methods which are frequently used in practice.

Just joining up the first and last points of a series will tend to be inaccurate as we are assuming that the residual in both cases will be zero, and this is very different from saying $E[e] = 0$. A simple way of improving upon this is to average the first few points of the series and average the same number of points at the end of the series, and then draw the line between the two average values. If we divide the data in two equal halves to do this the method is called 'semi-averages'. However, we should make it clear that we are not recommending these

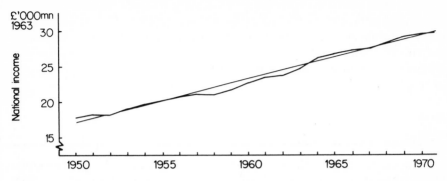

Figure 5.8   National Income (Net National Product at factor cost) United Kingdom 1950—71 (1963 prices). Source: *National Income and Expenditure* (Blue Book), H.M.S.O., 1972

methods to the reader but merely explaining them so he may understand them when he comes across their use by other writers.

A more important aspect to consider at this stage is to see whether the residual distances of the points from the line do in fact look random. If, as in Figure 5.8, most of the residuals at the beginning of the series are positive as are those at the end, and those in the middle are negative, then a slightly curved trend would represent the data better. However the advantage of the straight line is that we can draw it quickly and read off the value for the slope, b, by taking two years some distance apart, say 1955 and 1965, and reading off the trend values for national income from the line, £20,250 mn and £26,500, obtaining from these

$$b = \frac{Y_{1965} - Y_{1955}}{1965 - 1955} = 0.625 \tag{33}$$

Thus for every unit change in time, in this case a year, national income increased by £625 mn.

5.7.2 Constant Rate of Growth
In the same way that we can get a quick estimate of a linear relationship between a variable and time we can also measure its rate of change over time. Our hypothesis is now that a variable has a compound rate of growth, g (growth rather than contraction merely because this conforms to far more of our economic experience, but the hypothesis is equally valid for a negative rate). Thus we are saying that the variable increases by a fixed percentage of its value in the previous period, instead of the linear hypothesis that the variable increases by a fixed percentage of some base-year period. This is perhaps best expressed by the compound interest formula

$$Y_t = Y_0 (1 + g)^t \tag{34}$$

where the subscripts refer to time periods, 0 being the base or initial period of the series. The path of Y_t as t increases according to (34) is shown in Figure 5.9. If we let $Y_0 = a$ and $(1 + g) = b$ we have a comparable expression to the linear hypothesis

$$Y_t = ab^t \qquad (35)$$

In fact if we take logarithms of both sides,

$$\log Y_t = \log a + t \log b \qquad (36)$$

we can see from (36) that (35) is linear in logarithms. Thus if we graph $\log Y_t$ against t the curve set out in Figure 5.9 will now look like that shown in Figure 5.10.

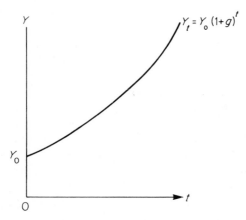

Figure 5.9   The constant rate of growth path

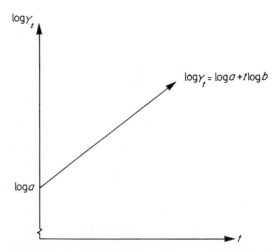

Figure 5.10   Constant rate of growth using semi-logarithmic axes

Before we consider an economic example we must introduce the stochastic element into the relationship. Thus (36) becomes

$$\log Y_t = \log a + t \log b + e \tag{37}$$

or setting $a_1 = \log a$ and $b_1 = \log b$

$$\log Y_t = a_1 + b_1 t + e \tag{38}$$

The choice of economic examples offers wide scope, since almost any variable which is expressed in money terms will have a strong rate of growth element in it as a result of inflation, even if it is not increasing in real terms. Perhaps the simplest example to take would be to look at National Income in current prices instead of constant prices. We have already seen that we can represent the constant price series by a linear trend. If we look at Figure 5.11 we can see that it might not be unreasonable to suggest that we can represent the current price series by a constant rate of growth trend of the form of (38). We can only attempt to fit the trend to the data by eye if we deal with a linear relationship, so the first step is to take logarithms of National Income. The resulting semi-logarithmic graph is shown in Figure 5.12. Note firstly that actual values are shown on the axes, not their logarithms, and secondly that the easiest way to draw such a graph is not to labour through the calculation of logarithms but to use a slide-rule instead of a ruler to measure the position of points in the vertical dimension, as slide-rules have a logarithmic scale (of course it is possible to obtain graph paper printed with a linear axis and a logarithmic axis).

Since we are fitting a straight line to data we can use exactly the same *ad hoc* methods for fitting (38) as for (32), remembering all the time that our variable is in logarithms. Hence if we add three numbers together and divide by three, we are finding the geometric mean of the

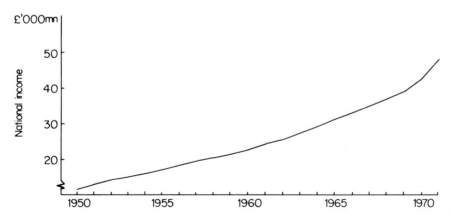

Figure 5.11   National Income at current prices, U.K. 1950—71. Source: *National Income and Expenditure*, 1972.

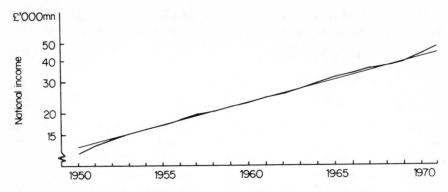

Figure 5.12   National Income at current prices, U.K. 1950—71 (using semi-logarithmic axes). Source: *National Income and Expenditure*, 1972

original data: because adding logarithms is equivalent to multiplication and dividing by three is equivalent to taking the cubic root. We can therefore also use ordinary least-squares regression to obtain a much more satisfactory estimate of the line.

Having plotted our line we can find its scope, b_1, by calculating as before

$$b_1 = \frac{\log Y_{1965} - \log Y_{1955}}{1965 - 1955} \tag{39}$$

However, b_1 is not the rate of growth of National Income

$$b_1 = \log b = \log(1 + g)$$

Hence

$$g = \text{antilog } b_1 - 1 \tag{40}$$

Using the data in Figure 5.12 we can substitute into (39) as follows

$$b_1 = \frac{\log(31{,}500) - \log(17{,}000)}{10} = \frac{4.4983 - 4.2305}{10} = 0.0268$$

Hence from (40) the rate of growth

$$g = \text{antilog } b_1 - 1 = 0.0637$$

or 6.37% per year.

5.8 Moving Averages

Thus far we have dealt with the situation where we impose a trend of a predetermined shape on the data. The second type of approach which is frequently used is to make an assumption about the nature of the fluctuations and see what the resulting trend looks like. This method is used when we have no reason to expect that the trend conforms to any simple function. The simplest example of this is when we feel there may be some cyclical pattern to the data. Again moving averages is an *ad hoc* method; the best method for the analysis of cyclical data is called spectral analysis. Unfortunately this method is rather complex, and is beyond the scope of this book as it is not often used in economic literature.

The assumption which is made about the nature of the fluctuations is that over some time period they will tend to average out to zero. Thus if we take the average of a set of observations which are in sequence over time we obtain a trend value. The problem is to choose the number of time periods to use in calculating the average. Sometimes we have a good reason for taking a particular average, for example, with quarterly data it is reasonable to suggest that we take a four period cycle. However, moving averages are normally used without any *a priori* knowledge and the final length of average decided upon purely from the look of the trend. It is thus possible for two people to choose different averages and for neither of them to be 'wrong'. The method is best shown by example.

The value of the physical increase in stocks and work in progress acts as an indicator of the relationship between producers' expectations of aggregate demand and the realized aggregate demand. If producers expect an increase in demand in the future they will increase their stocks so that they can meet that demand and vice-versa if they expect a decline. Secondly, if actual demand is less than that expected by producers stocks of unsold goods will be built up, and again stocks will decline if producers underestimate. We have a two-fold relationship which will affect the value of the physical increase in stocks and work in progress over time. Clearly we expect the variable to fluctuate and to be related to economic cycles in the economy as a whole. However, the nature of the exact relationship is by no means clear. A look at Figure 5.13 will show that over the period 1952 to 1971 the value of the physical increase in stocks and work in progress (in constant prices to avoid the complications of inflation) fluctuated widely. We now wish to know whether this was totally random, or if we can distinguish any trend.

The method of moving averages makes the assumption that the trend value for a particular period is an average of the period itself and those periods immediately before and after it. Let us call the original series W and the trend T. In the simplest case, for a three-period moving average we will calculate the trend value for period t as

$$T_t = \frac{W_{t-1} + W_t + W_{t+1}}{3} \tag{41}$$

If we consider that the two periods next closest in time are also relevant the five-period moving average value becomes

$$T_t = \frac{W_{t-2} + W_{t-1} + W_t + W_{t+1} + W_{t+2}}{5} \tag{42}$$

and so on for higher odd numbers. If we wish to take an even number of periods for our average, we have a problem since the average value will be calculated for an intermediate period between two years. Thus

$$\frac{W_t + W_{t+1} + W_{t+2} + W_{t+3}}{4}$$

is not equal to T_{t+1} or T_{t+2} but $T_{t+1.5}$. The method which is used to overcome this is to make half the transition from (41) to (42) by only assigning a weight of one-half to W_{t-2} and W_{t+2},

$$T_t = \frac{\frac{1}{2}W_{t-2} + W_{t-1} + W_t + W_{t+1} + \frac{1}{2}W_{t+2}}{4} \tag{43}$$

and the result is called a centred four-period moving average. In practice this is usually calculated as

$$T_t = \frac{W_{t-2} + 2W_{t-1} + 2W_t + 2W_{t+1} + W_{t+2}}{8} \tag{44}$$

but the result is clearly identical.

Table 5.2
Moving averages of the value of physical increase in stocks and work in progress
1952 to 1971

Year	(1) W	(2) Σ3	(3) Σ3/3	(4) Σ4	(5) Σ8	(6) Σ8/8	(7) Σ7	(8) Σ7/7
1952	63							
1953	130	245	81.7					
1954	52	484	161.3	547	1,266	158.3		
1955	302	589	196.3	719	1,549	193.6	1,130	161.4
1956	235	778	259.3	830	1,715	214.4	1,255	179.3
1957	241	583	194.3	885	1,656	207.0	1,753	250.4
1958	107	536	178.7	771	1,935	241.9	2,028	289.7
1959	188	923	307.7	1,164	2,414	301.8	1,792	256.0
1960	628	1,143	381.0	1,250	2,459	307.4	1,748	249.7
1961	327	1,021	340.3	1,209	2,421	302.6	2,140	305.7
1962	66	584	194.7	1,212	2,429	303.6	2,412	344.6
1963	191	890	296.7	1,217	2,486	310.8	2,474	353.4
1964	633	1,203	401.0	1,269	2,722	340.3	2,032	290.3
1965	379	1,262	420.7	1,453	2,901	362.6	1,866	266.6
1966	250	815	271.7	1,448	2,424	303.0	2,142	306.0
1967	186	597	199.0	976	1,915	239.4	2,216	316.6
1968	161	689	229.7	939	1,893	236.6	1,607	229.6
1969	342	768	256.0	954	1,746	218.3		
1970	265	631	210.3	792				
1971	24							

The averages are 'moving' because we calculate a value for each time period. Table 5.2 shows the original series as column (1), the moving sums of each set of three observations, $W_{t-1} + W_t + W_{t+1}$, as column (2) and the three-period moving average trend as column (3). We have no trend value for 1952 as we do not know W_{t-1} for that year and similarly no value for 1971 as we do not know W_{t+1}. Thus the more periods in our moving average the more observations we lose at each end of the series. Columns (7) and (8) give the values of the moving sums of seven observations and the seven-period moving average. The four-period moving average, column (6), is calculated firstly by adding up sums of four observations, column (4), and then adding these sums together in pairs, column (5). It is clear that the elements of column (5)

$$(W_{t-2} + W_{t-1} + W_t + W_{t+1}) + (W_{t-1} + W_t + W_{t+1} + W_{t+2})$$

give us the numerator of the right-hand side in (44). Hence column (6) is calculated by dividing column (5) by eight. These calculations are very simple, but if the 'moving sums' are each calculated from the previous one by subtracting the observation not required and adding

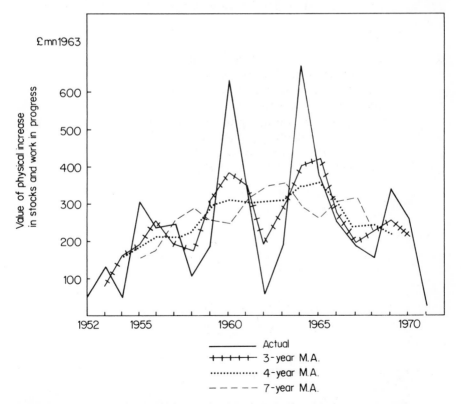

Figure 5.13   Value of physical increase in stocks and work in progress (1963 prices). Source: *National Income and Expenditure*, 1972

the new observation, any arithmetic slip will be carried on right down the column. It is, therefore, worth checking the last sum to make sure that this has not occurred.

Having calculated the three moving averages, we have graphed them on Figure 5.13. All three averages are smoother than the original series. The three-year moving average has the same pattern as the original, but has less pronounced fluctuations. The four-year moving average is smoother still and suggests a slow upward trend until 1965 when the series turns sharply downward. The seven-year moving average on the other hand fluctuates more than the four-year and bears much less relation to the original data. It has peaks in 1958, 1963 and 1967 where the actual data exhibit the opposite general pattern. It is in fact introducing a new cyclical pattern which is totally out of phase with the three-year moving average. This illustrates a very important danger in estimating moving averages, namely that there is a good chance that a new cycle will be introduced into the data purely as a result of the averaging process. Since moving averages were used originally to show up underlying cycles in economic series this presents a serious drawback. The trend calculated by moving averages should be used with great caution.

5.9 Calculating the Correlation Coefficient

In Table 5.1 we set out the components of the covariance and the correlation coefficient; however, if we wished to calculate these values we could use computationally easier procedures. We have already suggested (p. 18) that it is both quicker and more accurate to calculate the variance without calculating the mean of a variable first. We can proceed in exactly the same way with the covariance. The covariance between two variables X and Y, was defined as

$$\text{Cov}(X, Y) = \frac{\Sigma(X - \bar{X})(Y - \bar{Y})}{N}$$

Expanding

$$\frac{\Sigma(X - \bar{X})(Y - \bar{Y})}{N} = \frac{\Sigma XY}{N} - \frac{\Sigma X\bar{Y}}{N} - \frac{\Sigma Y\bar{X}}{N} + \frac{\Sigma \bar{X}\bar{Y}}{N}$$

Since $\bar{X}$ and $\bar{Y}$ are constant for all values of X and Y in the sample

$$\text{Cov}(X, Y) = \frac{\Sigma XY}{N} - \bar{Y}\frac{\Sigma X}{N} - \bar{X}\frac{\Sigma Y}{N} + \bar{X}\bar{Y}$$

but $\Sigma X/N = \bar{X}$ and $\Sigma Y/N = \bar{Y}$, hence

$$\text{Cov}(X, Y) = \frac{\Sigma XY}{N} - \bar{X}\bar{Y} - \bar{X}\bar{Y} + \bar{X}\bar{Y} = \frac{\Sigma XY}{N} - \bar{X}\bar{Y}$$

We can thus deal simply in terms of X and Y without bothering about deviations.

(i) Using a calculating machine   As many machines will automatically accumulate sums and sums of squares, calculate columns of

$$X \text{ and } X^2, \ Y \text{ and } Y^2, XY$$

then proceed to

$$\overline{X} = \Sigma X/N, \ \overline{Y} = \Sigma Y/N$$

$$\text{Var}(X) = \Sigma X^2/N - \overline{X}^2, \ \text{Var}(Y) = \Sigma Y^2/N - \overline{Y}^2$$

$$\text{Cov}(X, Y) = \Sigma XY/N - \overline{X}\,\overline{Y}$$

$$r = \frac{\text{Cov}(X, Y)}{\sqrt{[\text{Var}(X)\text{Var}(Y)]}}$$

(ii) Using a Computer   We can construct a program from the following flow chart.

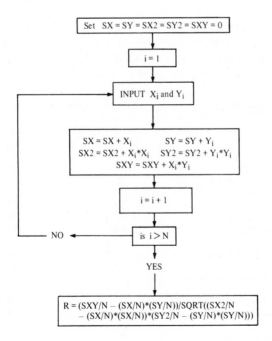

References and Suggested Reading

1.   We have introduced further important source material in this chapter:

Economic Trends, H.M.S.O., monthly.

National Income and Expenditure, H.M.S.O., annually (often referred to as the Blue Book).

Balance of Payments, H.M.S.O., annually (often referred to as the Pink Book, although a change of cover design in 1971 rendered this less appropriate).

The first of these is the main source of quarterly economic data for the U.K. It also contains a number of articles on the process of collection and publication of official data, which greatly aid their comprehension. In the October issue helpful supplementary tables are published. The Blue Book is the most widely used source of macro-economic data. The Pink Book provides a summary of the United Kingdom's transactions with foreign countries (and is referred to in question 4 and not in the text of the chapter).

2   R. A. Fisher, 'Frequency-distribution of the values of the correlation coefficient in samples from a normal bivariate population', *Biometrika*, 10, 507, 1915.

Questions

1   Using the following table estimate the correlation coefficient between the average weekly earnings of men aged 21 and over and the average weekly earnings of women aged 18 and over (full-time manual workers) in manufacturing industries in 1972.

Average Earnings of Manual Workers in certain industries, 1972

Industry	Men aged 21 and over	Women aged 18 and over
Food, drink and tobacco	35.8	17.3
Coal and petroleum products	38.9	20.5
Chemical and allied industries	36.8	18.6
Metal manufacture	38.0	18.8
Mechanical engineering	34.7	20.4
Instrument engineering	32.2	18.0
Electrical engineering	34.5	19.3
Shipbuilding and marine engineering	35.0	18.3
Vehicles	41.6	23.8
Metal goods not elsewhere specified	34.0	17.9
Textiles	32.1	17.3
Clothing and footwear	29.5	16.6
Bricks, pottery, glass, cement etc.	37.3	18.3
Timber, furniture etc.	34.1	19.7
Paper, printing and publishing	41.2	19.9

Source: *New Earnings Survey*

Test the hypothesis that there is no linear relationship between average earnings of the two groups of manual workers in these manufacturing industries in 1972.

Are there any drawbacks to using the correlation coefficient in this instance as a test statistic?

Comment on your results.

2 The table below is drawn from the 1969 *Family Expenditure Survey*

(i) Formulate an economic hypothesis for the relationship between household income and household expenditure on food.

(ii) Calcuate the correlation coefficient between weekly income and average weekly expenditure on food from the table.

(iii) You are told that in 1972 the correlation coefficient between household weekly income and average weekly expenditure on food was 0.95. Would you say that people's behaviour was the same in 1969 as 1972 on the basis of your sample?

Household weekly income and expenditure, 1969

Number of households 100s	Weekly income	Average weekly expenditure on food £
4	£6 and under £8	2.4
3	£8 and under £10	3.3
6	£10 and under £15	4.2
6	£15 and under £20	5.3
8	£20 and under £25	6.0
8	£25 and under £30	6.9
8	£30 and under £35	7.4
7	£35 and under £40	7.9
9	£40 and under £50	8.6
5	£50 and under £60	9.7

Source: *Family Expenditure Survey*

3 (i) Would you expect that the net capital stock of the U.K. at current replacement cost was best described by a linear trend or by a constant rate of growth?

(ii) Using the data set out below for the years 1960 to 1971 estimate these two trends by eye.

(iii) Can you say whether your expectations are fulfilled?

Net Capital Stock at Current Replacement Cost (£'000 million)

1960	1961	1962	1963	1964	1965	1966	1967	1968	1969	1970	1971
53.0	57.1	61.3	65.5	70.7	76.8	82.5	88.6	96.6	107.3	120.8	135.2

Source: *National Income and Expenditure*, 1972

4 Estimate a suitable trend for the balance of payments on current account over the period 1950–1971.

Justify your answer.

Balance of Payments on Current Account U.K. 1950—71 (£mn)

1950	1951	1952	1953	1954	1955	1956	1957	1958
307	−369	163	145	117	−155	208	233	344

1959	1960	1961	1962	1963	1964	1965	1966	1967
152	−255	6	122	124	−382	−49	84	−315

1968	1969	1970	1971
−271	444	681	1040

Source: *Balance of Payments* (Pink Book)

Answers
1   Let

X = average earnings of manual workers (men aged 21 and over)
Y = average earnings of manual workers (women aged 18 and over)

We wish to calculate

$$r_{XY} = \frac{\Sigma XY - \Sigma X \Sigma Y/N}{\sqrt{[\Sigma X^2 - (\Sigma X)^2/N][\Sigma Y^2 - (\Sigma Y)^2/N]}}$$

Setting this out in tabular form ($N = 15$):

	$X\,(£)$	$Y\,(£)$	X^2	Y^2	XY
	35.8	17.3	1,281.6	299.3	619.3
	38.9	20.5	1,513.2	420.3	797.5
	36.8	18.6	1,354.2	346.0	684.5
	38.0	18.8	1,444.0	353.4	714.4
	34.7	20.4	1,204.1	416.2	707.9
	32.2	18.0	1,036.8	324.0	579.6
	34.5	19.3	1,190.3	372.5	665.9
	35.0	18.3	1,225.0	334.9	640.5
	41.6	23.8	1,730.6	566.4	990.1
	34.0	17.9	1,156.0	320.4	608.6
	32.1	17.3	1,030.4	299.3	555.3
	29.5	16.6	870.3	275.6	489.7
	37.3	18.3	1,391.3	334.9	682.6
	34.1	19.7	1,162.8	388.1	671.8
	41.2	19.9	1,697.4	396.0	819.8
Σ	535.7	284.7	19,288.0	5,447.3	10,227.5
Σ/N	35.71	18.98	1,285.9	363.2	681.8

$\Sigma XY - \Sigma X \Sigma Y/N = 59.91$

$\Sigma X^2 - (\Sigma X)^2/N = 156.37$

$\Sigma Y^2 - (\Sigma Y)^2/N = 43.69$

$r = 0.7248$

$H_0: \rho = 0, \quad H_1: \rho > 0, \quad$ Significance level chosen: 5%

$t(13)_{0.05} = 1.77$

$$t = \frac{r - \rho}{\sqrt{[(1 - r^2)/(n - 2)]}} = \frac{0.7248}{\sqrt{[(1 - 0.5253)/13]}} = 3.79$$

$t > t(13)_{0.05}$, so we reject H_0.

From these data we assert that there is a positive correlation between X and Y (given our 5% significance level).

Drawbacks
   (1) We have no indication of the number of wage earners in each industry. It is likely that it is different in each industry. Therefore, by giving an equal weight to each industry we may be biasing our estimate of r.
   (2) We have no specific hypothesis about the linear relationship between male and female average earnings. We know that there is a general tendency for women to be paid less than men, but the differential is caused by a large number of institutional factors which will be different in each industry and even within the industry. It is thus doubtful if we really want to say that there is a stochastic relationship

$$Y = a + bX + u$$

or

$$X = c + dY + v$$

for that matter. The correlation coefficient is rather more a descriptive statistic in this case.

Comments   The reader should remark on:
   (i) The size of the correlation coefficient.
   (ii) The fact that this implies that the higher male average earnings the greater the absolute differential between the two sexes.
   (iii) We have no information on whether women are paid less for the same work.
   (iv) The size of r makes it unlikely that any bias introduced by ignoring the different numbers of workers in each industry will affect the general conclusion.

2   (i) A simple consumption function for households might run

$$C = a + bY + u \tag{1}$$

where C is average weekly consumption of food,
         Y is weekly income.

However, no allowance has been made for differences in family size. We must assume that aggregating means that (1) is not affected by the size of the households or the age of their members.

Engel's Law might suggest that a linear function is not the right shape, but if a is positive and b is such that $0 < b < 1$, the proportion of income spent on food will fall as income rises.

(ii) We have a frequency distribution in this case, and must therefore set out our calculations as follows:

$$\text{Var}(C) = \frac{\Sigma f(C - \bar{C})^2}{\Sigma f} = \frac{\Sigma fC^2}{\Sigma f} - \left(\frac{\Sigma fC}{\Sigma f}\right)^2$$

$$\text{Var}(Y) = \frac{\Sigma f(Y - \bar{Y})^2}{\Sigma f} = \frac{\Sigma fY^2}{\Sigma f} - \left(\frac{\Sigma fY}{\Sigma f}\right)^2$$

$$\text{Cov}(C, Y) = \frac{\Sigma f(C - \bar{C})(Y - \bar{Y})}{\Sigma f} = \frac{\Sigma fCY}{\Sigma f} - \frac{\Sigma fC\Sigma fY}{(\Sigma f)^2}$$

$$r = \frac{\text{Cov}(C, Y)}{\sqrt{[\text{Var}(C)\text{Var}(Y)]}}$$

f	C	Midpoint Y	fC	fY	fC^2	fY^2	fCY
4	2.4	7	9.6	28	23.04	196	67.2
3	3.3	9	9.9	27	32.67	243	89.1
6	4.2	12.5	25.2	75	105.84	937.5	315.0
6	5.3	17.5	31.8	105	168.54	1,837.5	556.5
8	6.0	22.5	48.0	180	288.0	4,050	1,080.0
8	6.9	27.5	55.2	220	380.88	6,050	1,518.0
8	7.4	32.5	59.2	260	438.08	8,450	1,924.0
7	7.9	37.5	55.3	262.5	436.87	9,843.8	2,073.75
9	8.6	45	77.4	405	665.64	18,225	3,483.0
5	9.7	55	48.5	275	470.45	15,125	2,667.5
Σ 64			420.1	1,837.5	3,010.0	64,957.8	13,774.1

$$\text{Var}(C) = \frac{3,010}{64} - \left(\frac{420.1}{64}\right)^2 = 3.944$$

$$\text{Var}(Y) = \frac{64,958}{64} - \left(\frac{1,838}{64}\right)^2 = 190.2$$

$$\text{Cov}(C, Y) = \frac{13,774}{64} - \frac{(420.1)(1,838)}{64^2} = 26.71$$

$$r_{CY} = \frac{26.71}{\sqrt{[(3.944)(190.2)]}} = 0.975$$

(iii) Since we are comparing a sample of households with the actual value for 1972, our null hypothesis is

$$H_0: \rho = 0.95$$

leaving us with an alternative hypothesis of

$$H_1: \rho \neq 0.95$$

the test statistic we wish to calculate is

$$z^* = \frac{z - \zeta}{\sqrt{[1/(n-3)]}}$$

We know that z is approximately $\sim N(\zeta, 1/(n-3))$.
  Thus if we choose 5% as our significance level

$$z_{0.05} = 1.96$$

Transforming ρ

$$\zeta = \tfrac{1}{2} \log_e \left(\frac{1 + \rho}{1 - \rho} \right) = 1.832$$

Transforming r

$$z = \tfrac{1}{2} \log_e \left(\frac{1 + r}{1 - r} \right) = 2.185$$

$$z^* = \frac{2.185 - 1.832}{\sqrt{(1/6,397)}} = 28.3$$

Therefore as $z^* > z_{0.05}$ we reject the null hypothesis.

  Note that in the denominator of z^* we had to use the actual number of households and not the number of hundreds of households as before as it does not cancel on this occasion.

  (Problems arising from the grouping of the data have been ignored.)

3 (i) The value of net capital stock at current replacement cost is affected by 3 main factors:

  (a) current replacement cost,
  (b) gross fixed capital formation,
  (c) depreciation.

The whole variable can be described approximately by the following relationship

$$S_t = S_{t-1}(1 - d) \left(\frac{P_t}{P_{t-1}} \right) + I_t$$

where S is the net capital stock at current replacement cost,
  d is the rate of depreciation,
  P is the replacement cost,
  I is gross fixed capital formation and
  t denotes the year.

We revalue S_{t-1} in t's prices by multiplying by P_t/P_{t-1}, and calculate how much the stock has been reduced by depreciation by multiplying by $(1-d)$ and finally we must add in the new capital stock created during the year.

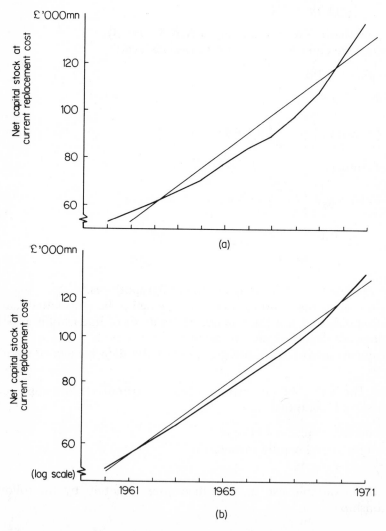

Figure 5.14   (a) Linear trend. (b) Constant rate of growth. Source: *National Income and Expenditure*, 1972

As a result of (a) we have inflation in the series, which leads us towards expecting a rate of growth rather than a linear trend.

(b) will flucutuate with the performance of the economy, which does not lead us to either hypothesis.

(c) is at a rate d fixed by convention.

(ii) Reading the values from Figure 5.14(a) we can substitute the appropriate values into equation (33) to determine the slope of the linear trend

$$b = \frac{S_{1969} - S_{1963}}{1969 - 1963} = \frac{108 - 66}{1969 - 1963} = \frac{42}{6} = 7$$

(the years 1969 and 1963 have been chosen as they are near the upper and lower ends of the period respectively).

In order to estimate a constant rate of growth we can use the line drawn in Figure 5.14(b), where the vertical axis is logarithmic, and substitute the values obtained from it in equation (39)

$$b_1 = \frac{\log S_{1969} - \log S_{1963}}{1969 - 1963} = \frac{\log 109 - \log 66}{1969 - 1963} = 0.0363$$

Thus the rate of growth = antilog $b_1 - 1 = 1.09 - 1 = 0.09$. I.e. the rate of growth is 9% p.a.

Since both trends were drawn by eye the estimated values of b and b_1 will vary with the choice of line.

(iii) Since the vertical scales of the two graphs are different it is not possible to make any direct comparison. However, the linear graph shows a pronounced curve to the original series indicating that a compound rate of growth hypothesis would be more appropriate than the linear trend. The semi-logarithmic graph also shows an upward curve in the original series indicating that the rate of growth increases with time. Thus while neither hypothesis is a very good approximation to the original series the constant rate of growth trend is a move in the right direction from a linear trend.

4   Clearly the first step is to draw a graph of the series, as in Figure 5.15(a). We can see immediately that neither a simple linear trend nor a constant rate of growth hypothesis are appropriate in this case. It is arguable that since devaluation in 1967 might be expected to have a substantial effect on any underlying trend we should consider the pre-1967 and the post-1967 figures separately. However, it is possible that the use of moving averages might help to show any underlying movement. We have no prior hypothesis about the most suitable length of average, so we shall have to judge our choice by result. The balance of payments as a whole (Net Currency Flow) obviously has a long-run value of zero but capital transactions can

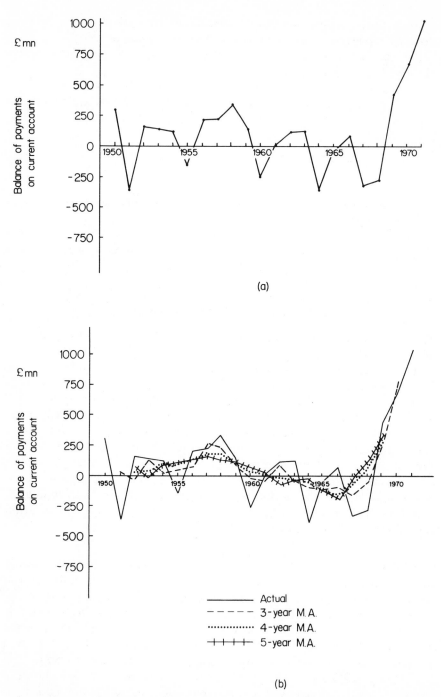

Figure 5.15 (a) U.K. balance of payments on current account, 1950—71. (b) Moving average trends in the U.K. balance of payments on current account, 1950—71. Source: *Balance of Payments*

balance out deficits/surpluses on Current Account for quite long periods. Thus although the trend may tend towards the horizontal it need not necessarily do so. However we would expect that the longer the trend we take the more it will tend to the horizontal.

Since we only have 22 data points a long average is clearly inappropriate. Taking the shorter moving averages of 3, 4 and 5 years, the smoothest of these is the 4-year average, the 3-year follows the original data more closely, and the 5-year average has more turning points. In accepting the 4-year average we are assuming that the underlying pattern is smooth.

Year	Original series	Σ 3	Σ 3/3	Σ 4	Σ 8	Σ 8/8	Σ 5	Σ 5/5
1950	307							
1951	−369	101	33.7					
				246				
1952	163	−61	−20.3		302	37.75	363	72.6
				56				
1953	145	425	141.7		326	40.75	−99	−19.8
				270				
1954	117	107	35.7		585	73.125	478	95.6
				315				
1955	−155	170	56.7		718	89.75	548	109.6
				403				
1956	208	286	95.3		1,033	129.125	747	149.4
				630				
1957	233	785	261.7		1,567	195.875	782	156.4
				937				
1958	344	729	243		1,411	176.375	682	136.4
				474				
1959	152	241	80.3		721	90.125	480	96
				247				
1960	−255	−97	−32.3		272	34	369	73.8
				25				
1961	6	−127	−42.3		22	2.75	149	29.8
				−3				
1962	122	252	84		−133	−16.625	−385	−77
				−130				
1963	124	−136	−45.3		−315	−39.375	−179	−35.8
				−185				
1964	−382	−307	−102.3		−408	−51	−101	−20.2
				−223				
1965	−49	−347	−115.7		−885	−110.625	−538	−107.6
				−662				
1966	84	−280	−93.3		−1,213	−151.625	−933	−186.6
				−551				
1967	−315	−502	−167.3		−609	−76.125	−107	−21.4
				−58				
1968	−271	−142	−47.3		481	60.125	623	124.6
				539				
1969	444	854	284.7		2,433	304.125	1,579	315.8
				1,894				
1970	681	2,165	721.7					
1971	1,040							

6 Regression

In the previous chapter we derived a statistic to show how close the linear relationship between two variables is. We also showed that we can describe any such linear relationship between two variables X and Y by the function

$$Y = a + bX \tag{1}$$

However, we also noted that many economic relationships are stochastic, i.e. for any particular values of Y and X say Y_i and X_i, (1) is actually of the form

$$Y_i = a + bX_i + e_i \tag{2}$$

The problem we shall consider in this chapter is how to estimate the values of a and b when, for any particular sample we can take, the relationship (1) is disturbed by a set of errors e.

6.1 Estimation
Let us consider the problem of estimating the relationship between income and consumption in the United Kingdom over the past few years. A simple macroeconomic consumption function might run

$$C = a + bY \tag{3}$$

where C stands for aggregate consumption and Y for aggregate income, b is the marginal propensity to consume, $0 < b < 1$, and a is a positive constant, so that as income rises the ratio of consumption to income, C/Y, the average propensity to consume, falls. (The diagrammatic form of (3) is shown in Figure 6.1.)

As we are dealing with measurable variables, or, more importantly, variables which actually have been measured in the U.K., we must define our terms more closely. Our basic hypothesis is that in aggregate

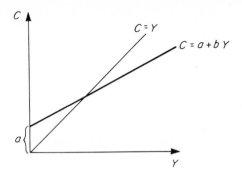

Figure 6.1

the amount that people spend on consumption is linearly determined by the amount they have to spend. So we can define consumption as consumers' expenditure. Since we are only dealing with people and not firms and the government as well, income must be personal income, but we must deduct direct taxation from this as that part of income cannot be spent on consumption, and correspondingly add in transfers to persons, (national insurance and social security benefits etc.). Thus our income variable is personal disposable income. Finally we must take into account that our variables have a different unit of measurement in each time period as prices have changed. If all prices double we would expect consumption to double (assuming that income also doubles), but if C and Y are measured in the prices prevailing in each time period (current prices), C will increase less than proportionately with increases in Y (because $a > 0$ and $0 < b < 1$). We must, therefore, measure both variables in constant units over all time periods. Ideally perhaps we could measure consumption in physical units, but this would be difficult firstly because we are talking about using an aggregate and would be adding together different sorts of items and secondly because the individual commodities may change. Let us, therefore, choose constant prices and measure each year's consumption and income in the prices of one particular year.

In Figure 6.2 we have plotted the values of C against the values of Y using quarterly data for the years 1968 to 1972 and the actual values are shown in the first two columns of Table 6.1. (We have obtained these data from *Economic Trends*, which is the most substantial source of quarterly economic data for the U.K.)

It is immediately obvious that the data conform quite closely to a straight line. We have to decide which straight line best represents the relationship between Y and C. In the last chapter we said that since (3) is a stochastic relationship we believe that

$$E[C|Y] = a + bY \tag{4}$$

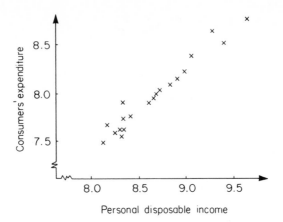

Figure 6.2 Consumers' expenditure and personal disposable income in the U.K., quarterly 1968—72, £'000mn 1970. Source: *Economic Trends*

and consequently

$$E[e] = 0 \qquad (5)$$

Our particular sample gives us 20 values of C and 20 values of Y from which we must estimate a and b. Given these estimates we can go on to calculate

$$\hat{C}_i = \hat{a} + \hat{b}Y_i \qquad (6)$$

where $\hat{}$ denotes an estimate for the $(i = 1, \ldots, 20)$ values of Y. Lastly we will obtain

$$C_i - \hat{C}_i = \hat{e}_i \qquad (7)$$

where the $\hat{e}_i$ are the residuals in the relationship. (Since our $\hat{a}$ and $\hat{b}$ are only estimates of a and b, the $\hat{e}_i$ we derive from (7) are not necessarily the random disturbances to (3).)

We must therefore derive some method of estimation of a and b for the model described in (4) which we feel is 'best'. Ordinary Least Squares is such a method, and we shall now set out its explanation.

6.2 Ordinary Least Squares
The method of ordinary least squares is so called because it estimates a and b in such a way that the sum of squared residuals, $\Sigma \hat{e}_i^2$, is as small as possible. We shall begin our exposition, however, by considering the properties of the e_i, the disturbances in (3), in more detail as they affect the sense in which ordinary least squares is 'best'.

We have assumed in (5) that these disturbances have a mean of zero; let us also make explicit assumptions about the exact nature of their

Table 6.1
The regression of consumption on income

t		C	Y	C^2	Y^2	CY	$\hat{C}$	$\hat{e}$	$\hat{e}^2$
1968	Q1	7.80	8.39	60.840	70.392	65.442	7.719	0.081	0.0066
	Q2	7.48	8.18	55.950	66.912	61.186	7.545	−0.065	0.0042
	Q3	7.60	8.25	57.760	68.063	62.700	7.603	−0.003	0.0000
	Q4	7.64	8.28	58.370	68.558	63.259	7.628	0.012	0.0001
1969	Q1	7.58	8.32	57.456	69.222	63.066	7.661	−0.081	0.0066
	Q2	7.66	8.20	58.676	67.240	62.812	7.562	0.098	0.0096
	Q3	7.63	8.29	58.217	68.724	63.253	7.636	−0.006	0.0000
	Q4	7.73	8.40	59.753	70.560	64.932	7.728	0.002	0.0000
1970	Q1	7.70	8.32	59.290	69.222	64.064	7.661	0.039	0.0015
	Q2	7.82	8.61	61.152	74.132	67.330	7.902	−0.082	0.0067
	Q3	7.91	8.71	62.568	75.864	68.896	7.985	−0.075	0.0056
	Q4	7.92	8.69	62.726	75.516	68.825	7.968	−0.048	0.0023
1971	Q1	7.86	8.67	61.780	75.169	68.146	7.951	−0.091	0.0083
	Q2	8.00	8.68	64.000	75.342	69.440	7.960	0.040	0.0016
	Q3	8.11	8.80	65.772	77.440	71.368	8.059	0.051	0.0026
	Q4	8.21	8.94	67.404	79.924	73.397	8.175	0.035	0.0012
1972	Q1	8.32	8.97	69.222	80.461	74.630	8.200	0.120	0.0144
	Q2	8.45	9.45	71.403	89.303	79.853	8.598	−0.148	0.0219
	Q3	8.58	9.31	73.616	86.676	79.880	8.482	0.098	0.0096
	Q4	8.73	9.58	76.213	91.776	83.633	8.706	0.024	0.0006
	Σ	158.73	173.04	1,262.168	1,500.496	1,376.112		0.001	0.1034
$N = 20$ Σ/N		7.9365	8.6520	63.1084	75.0248	68.8056		0.000	0.0057a

a $\Sigma/(N-2)$.

distribution, firstly that their distribution does not vary with i, so each e_i has the same variance, namely σ^2 which is unknown, secondly that the disturbances are independent of each other, i.e. $E[e_i e_j] = 0$ for all i and j, $i \neq j$, and lastly that they conform to a normal distribution, thus the larger (in absolute terms) the disturbance the less likely it is to occur. Therefore for each particular Y_i that we have in our sample we could, according to (4), (5) and the distribution of the disturbances, observe any one of an infinite number of possible C_i. If e_i for a given Y_i is $N(0, \sigma^2)$ then C_i for the same Y_i is $N(a + b Y_i, \sigma^2)$.

Using the $\hat{e}_i$ we calculate and these assumptions, we can derive estimators which are in some sense best. The least-squares estimators have the following properties: firstly $\Sigma \hat{e}_i = \bar{\hat{e}} = 0$ and secondly $\Sigma \hat{e}_i^2$ is a minimum. Thus we are imposing some of the features of the disturbances on the errors in our sample. The choice of $\Sigma \hat{e}_i = 0$ is obviously attractive, and minimizing the variance of the residuals tries to make the line fit as well as possible. We estimate a and b as

$$\hat{b} = \frac{\text{Cov}(C, Y)}{\text{Var } Y} \tag{8}$$

$$\hat{a} = \bar{C} - \hat{b} \bar{Y} \tag{9}$$

If the reader can follow the calculus we can derive these estimators easily. In any case the reader can see that, if we have calculated $\hat{b}$, since $\hat{e} = 0$ the line must pass through $(\bar{Y}, \bar{C})$. Hence

$$\bar{C} = \hat{a} + \hat{b} \bar{Y} \tag{10}$$

and we can estimate $\hat{a}$ from (9) as $\bar{C}$, $\hat{b}$ and $\bar{Y}$ are known.

We said that $\Sigma \hat{e}_i^2$ should be a minimum. We know that

$$\Sigma \hat{e}_i^2 = \Sigma (C_i - \hat{a} - \hat{b} Y_i)^2 \tag{11}$$

and we must select suitable values for $\hat{a}$ and $\hat{b}$. The process of minimizing $\Sigma \hat{e}_i^2$ with respect to $\hat{a}$ and $\hat{b}$ will result in obtaining two equations which we can solve for $\hat{a}$ and $\hat{b}$.

★ We must, therefore, find the values where
★
★ $$\frac{\partial \Sigma \hat{e}_i^2}{\partial \hat{a}} = \frac{\partial \Sigma \hat{e}_i^2}{\partial \hat{b}} = 0$$
★
★ and the appropriate second-order conditions for a minimum hold.
★
★ $$\frac{\partial \Sigma \hat{e}_i^2}{\partial \hat{a}} = 2\Sigma (C_i - \hat{a} - \hat{b} Y_i)(-1) = -2\Sigma (C_i - \hat{a} - \hat{b} Y_i) \tag{12}$$
★

$$\frac{\partial \Sigma \hat{e}_i^2}{\partial \hat{b}} = 2\Sigma(C_i - \hat{a} - \hat{b}Y_i)(-Y_i) = -2\Sigma(C_i Y_i - \hat{a}Y_i - \hat{b}Y_i^2)$$

(13)

Setting both equations equal to zero and dividing both sides by 2 gives us

$$\Sigma(C_i - \hat{a} - \hat{b}Y_i) = 0 \qquad (14)$$

$$\Sigma(C_i Y_i - \hat{a}Y_i - \hat{b}Y_i^2) = 0 \qquad (15)$$

If N is the size of the sample the two equations we obtain from rearranging (14) and (15) are

$$N\hat{a} + \hat{b}\Sigma Y_i = \Sigma C_i \qquad (16)$$

$$\hat{a}\Sigma Y_i + \hat{b}\Sigma Y_i^2 = \Sigma C_i Y_i \qquad (17)$$

Equations (16) and (17) are known as the *Normal Equations*.

Solving these, let us multiply (16) by ΣY_i and divide by N

$$\hat{a}\Sigma Y_i + \hat{b}(\Sigma Y_i)^2 / N = \Sigma C_i \Sigma Y_i / N \qquad (18)$$

Subtracting (18) from (17)

$$\hat{b}[\Sigma Y_i^2 - (\Sigma Y_i)^2 / N] = \Sigma C_i Y_i - \Sigma C_i \Sigma Y_i / N \qquad (19)$$

and simplifying

$$\hat{b} = \frac{\Sigma C_i Y_i - \Sigma C_i \Sigma Y_i / N}{\Sigma Y_i^2 - (\Sigma Y_i)^2 / N} \qquad (20)$$

However, if we look at the right-hand side of (20) we can see that the numerator is in fact $N \, \text{Cov}(C, Y)$ and the denominator $N \, \text{Var} \, Y$ (see p. 99 for their definition) and thus we have derived (8).

(Lastly it is easy to check that the three second-order conditions for a minimum hold in this case.)

We now have to estimate $\hat{a}$ and $\hat{b}$ for our particular example. Clearly we must calculate $\hat{b}$ first from (20) and $\hat{a}$ second from (9). In Table 6.1 we have set out how we would do this using a calculating machine. (Those wishing to calculate the statistics by hand should consult p. 20.)

$$\hat{b} = \frac{68.8056 - (7.9365)(8.652)}{75.0248 - 8.652^2} = 0.829$$

$$\hat{a} = 7.937 - (0.829)(8.652) = 0.764$$

So the equation of our line is $C = 0.764 + 0.829\,Y$ and our estimates of a and b fulfil our expectations about their sign and size.

6.3 Closeness of Association

In Chapter 5 we covered the closeness of association between two variables. We could in the same way calculate the correlation coefficient between consumption and income. Let us recall that

$$r = \frac{\text{Cov}(C,\ Y)}{\sqrt{(\text{Var}\ C\ \text{Var}\ Y)}} \tag{21}$$

Substituting in from Table 6.1

$$r = \frac{68.8056 - (7.9365)(8.652)}{\sqrt{[(63.1084 - 7.9365^2)(75.0248 - 8.652^2)]}} = 0.978$$

We can see that the relationship is quite close, and could of course go on to test the hypothesis that $r = 0$.

In regression analysis r^2 is often quoted rather than r as a measure of the closeness of association of the two variables as r^2 can be interpreted as the ratio of explained to total variation in C. Let us show this:

$$r^2 = \frac{[\text{Cov}(C,\ Y)]^2}{\text{Var}\ C\ \text{Var}\ Y}$$

$$= \frac{\hat{b}^2\ \text{Var}\ Y}{\text{Var}\ C}$$

multiplying the numerator and denominator by N

$$r^2 = \frac{\hat{b}^2\ \Sigma(Y_i - \overline{Y})^2}{\Sigma(C_i - \overline{C})^2} \tag{22}$$

The denominator of (22) is clearly total variation in C and we can show that the numerator is the 'explained' variation. From (6), (7) and (9)

$$C_i - \overline{C} = \hat{b}(Y_i - \overline{Y}) + \hat{e}_i \tag{23}$$

Hence squaring both sides and summing over N

$$\Sigma(C_i - \overline{C})^2 = \hat{b}^2\,\Sigma(Y_i - Y)^2 + \Sigma\hat{e}_i^2 + 2\hat{b}\Sigma(Y_i - \overline{Y})\hat{e}_i$$

$$= \hat{b}^2\,\Sigma(Y_i - \overline{Y})^2 + \Sigma\hat{e}_i^2 \tag{24}$$

as

$$\Sigma(Y_i - \overline{Y})\hat{e}_i = 0 \tag{25}$$

$\Sigma\hat{e}_i^2$ is clearly the variation in C which is not explained by the regression and thus $\hat{b}^2\,\Sigma(Y_i - \overline{Y})^2$ is the explained variation. However, equation (25) is not an obvious truth, but we can show it if we substitute in $\Sigma(Y_i - \overline{Y})\hat{e}_i$ for $\hat{e}_i$ from (23).

$$\Sigma(Y_i - \bar{Y})\hat{e}_i = \Sigma(Y_i - \bar{Y})[(C_i - \bar{C}) - \hat{b}(Y_i - \bar{Y})]$$
$$= \Sigma(Y_i - \bar{Y})(C_i - \bar{C}) - \hat{b}\Sigma(Y_i - \bar{Y})^2 \qquad (26)$$

but from (20) we know that

$$\hat{b} = \frac{\Sigma(Y_i - \bar{Y})(C_i - \bar{C})}{\Sigma(Y_i - \bar{Y})^2} \qquad (27)$$

so

$$\hat{b}\Sigma(Y_i - \bar{Y})^2 = \Sigma(Y_i - \bar{Y})(C_i - \bar{C})$$

and thus

$$\Sigma(Y_i - \bar{Y})\hat{e}_i = 0$$

Therefore, in our example,

$$r^2 = 0.978^2 = 0.956$$

and thus the regression $C = 0.764 + 0.829Y$ explains 96% of the variation in C.

6.4 Properties of Least-Squares Estimators

In introducing the estimators of a and b which we have used we averred that they were in some way preferable to other estimators. Obviously they are far less arbitrary than just drawing a line through the points by eye. Firstly, they are unbiased estimators, i.e.

$$E[\hat{b}] = b \qquad (28)$$

and

$$E[\hat{a}] = a \qquad (29)$$

Secondly, they are 'best' in the sense that they have the minimum variance of all possible linear unbiased estimators. They are referred to as being BLUE, *Best Linear Unbiased Estimators*.

6.4.1 Linearity

We have not referred to the 'linear' element in BLUE. Both $\hat{a}$ and $\hat{b}$ are linear functions of C_i. Let us consider $\hat{b}$ first. If we decompose (20)

$$\hat{b} = \Sigma \left\{ \left[\frac{Y_i - \Sigma Y_i/N}{\Sigma Y_i^2 - (\Sigma Y_i)^2/N} \right] C_i \right\} \qquad (30)$$
$$= \Sigma A_i C_i$$

where

$$A_i = \frac{Y_i - \Sigma Y_i/N}{\Sigma Y_i^2 - (\Sigma Y_i)^2/N} = \frac{Y_i - \bar{Y}}{\Sigma(Y_i - \bar{Y})^2}$$

Similarly from (9) we can express $\hat{a}$ as a linear function of C_i

$$\hat{a} = \Sigma C_i/N - \hat{b}\overline{Y} \tag{31}$$

Substituting from (30)

$$\hat{a} = \Sigma C_i/N - \Sigma A_i C_i \overline{Y} \tag{32}$$

Factorizing

$$\hat{a} = \Sigma[(1/N - A_i \overline{Y})C_i] \tag{33}$$

6.4.2 Unbiasedness

Unbiasedness is also easy to show. If we substitute for C_i in (30) using the relevant interpretation of (2)

$$\hat{b} = \Sigma A_i(a + bY_i + e_i) \tag{34}$$

we can show that

$$\hat{b} = b + \Sigma A_i e_i \tag{35}$$

because $\Sigma A_i = 0$ as $\Sigma(Y_i - \overline{Y}) = 0$ and hence $\Sigma A_i a = 0$; further $\Sigma A_i Y_i = 1$ as $\Sigma(Y_i - \overline{Y})Y_i = \Sigma Y_i^2 - (\Sigma Y_i)^2/N$ and hence $\Sigma A_i bY_i = b$. Taking expectations

$$E[\hat{b}] = b + \Sigma A_i E[e_i] \tag{36}$$

but we know $E[e_i] = 0$, so

$$E[\hat{b}] = b$$

and $\hat{b}$ is an unbiased estimator of b. We can proceed in the same way for $\hat{a}$ by substituting into (33) for C_i from the appropriate version of (2).

6.4.3 Minimum Variance

The minimum variance property is slightly more difficult to grasp. We must firstly determine the variance of our estimators, and then derive an expression for all other linear unbiased estimators of b, let us say $\hat{b}_1$, where $\hat{b}_1 = \Sigma B_i C_i$, $B_i = A_i + D_i$ and the D_i are any arbitrary set of constants such that $\hat{b}_1$ is unbiased. We can then derive the variance of $\hat{b}_1$, and show that this variance contains a non-negative term in addition to the variance of $\hat{b}$, which is only zero if all the D_i are zero. Again we can show the same sequence for $\hat{a}$. If the reader wishes to check this he is referred to J.Johnston, *Econometric Methods*, McGraw—Hill.

We can thus see why the least-squares estimators are so widely used. Obviously if we have information about a and b other than that in the sample we may be able to obtain a better estimator by taking this information into account. However, given our assumptions about the model and the disturbances our least-squares estimators are the BLUEs.

6.5 Testing the Coefficients

Since we are sampling, our least-squares estimators are distributed. The particular set of observations that we can have on income and consumption are only a few that we can obtain from the population. We already know that $E[\hat{b}] = b$; we now need to show how $\hat{b}$ is distributed round its mean. The deviation of any individual $\hat{b}$ from its mean is $\hat{b} - b$, and we know from (35) that

$$\hat{b} - b = \Sigma A_i e_i \tag{37}$$

Thus

$$\text{Var}(\hat{b}) = E[\hat{b} - b]^2$$
$$= E[\Sigma A_i e_i]^2 \tag{38}$$

Decomposing

$$\text{Var}(\hat{b}) = E[A_1^2 e_1^2 + A_1 A_2 e_1 e_2 + \ldots + A_1 A_N e_1 e_N + A_2 A_1 e_2 e_1$$
$$+ A_2^2 e_2^2 + \ldots + A_N^2 e_N^2] \tag{39}$$

We assumed originally that $E[e_i^2] = \sigma_e^2$ and $E[e_i e_j] = 0$ $(i \neq j)$, thus (39) becomes

$$\text{Var}(\hat{b}) = A_1^2 \sigma_e^2 + A_2^2 \sigma_e^2 + \ldots + A_N^2 \sigma_e^2$$
$$= \sigma_e^2 \Sigma A_i^2 \tag{40}$$

However, we have defined $A_i = (Y_i - \overline{Y})/\Sigma(Y_i - \overline{Y})^2$ and hence $\Sigma A_i^2 = 1/\Sigma(Y_i - \overline{Y})^2$, so we can say

$$\text{Var}(\hat{b}) = \sigma_e^2 / \Sigma(Y_i - \overline{Y})^2 \tag{41}$$

Similarly we can show that

$$\text{Var}(\hat{a}) = \sigma_e^2 \left[\frac{1}{N} + \frac{\overline{Y}^2}{\Sigma(Y_i - \overline{Y})^2} \right] \tag{42}$$

Since we know the means and variances of our estimators, we can now derive suitable test statistics for them under the null hypotheses of $a = 0$ and $b = 0$. In our example we would be testing to see if consumption was not significantly different from zero when income was zero, and secondly if there was a relationship between consumption and income. It is important to realize that testing $H_0 : b = 0$ is exactly equivalent to testing that $H_0 : r = 0$. As we found before (p. 48), the distribution of the statistics will depend upon whether σ_e^2 is known or unknown, as both $\text{Var}(\hat{a})$ and $\text{Var}(\hat{b})$ depend upon σ_e^2. From p. 114 we can say $\hat{a}$ is $N(a, \text{Var}(\hat{a}))$ and $\hat{b}$ is $N(b, \text{Var}(\hat{b}))$. Thus if we standardize, our test statistics are

$$Z_a = \frac{(\hat{a} - a)\sqrt{[N\Sigma(Y_i - \overline{Y})^2]}}{\sigma_e \sqrt{\Sigma Y_i^2}} \tag{43}$$

$$Z_b = \frac{(\hat{b} - b)\sqrt{[\Sigma(Y_i - \bar{Y})^2]}}{\sigma_e} \tag{44}$$

But if σ_e is unknown we must estimate it from our sample using

$$\hat{\sigma}_e^2 = \frac{\Sigma\hat{e}_i^2}{(N - 2)}$$

where our unbiased estimator allows for the loss of two degrees of freedom. In this case our test statistic will be

$$t_a = \frac{(\hat{a} - a)\sqrt{[N\Sigma(Y_i - \bar{Y})^2]}}{\hat{\sigma}_e\sqrt{\Sigma Y^2}} \tag{45}$$

$$t_b = \frac{(\hat{b} - b)\sqrt{[\Sigma(Y_i - \bar{Y})^2]}}{\hat{\sigma}_e} \tag{46}$$

both distributed as t with $N - 2$ degrees of freedom.

For each of our regression coefficients let us test the hypothesis that it is not significantly different from zero. We shall impose a harsh criterion as we are dealing with a very basic hypothesis and say that the significance level shall be 1%. As $N = 20$ we have 18 degrees of freedom. This is a two-tailed test as in the first case $H_0: a = 0$ and consequently under $H_1: a \neq 0$ and in the second we have $H_0: b = 0$ and $H_1: b \neq 0$. Thus if we look up the critical value of t in Appendix Table A3.3, we can see that we should accept H_0 if

$$-2.88 \leqslant t \leqslant 2.88$$

When we calculate t_a and t_b ($\hat{\sigma}_e^2 = 0.0057$, $\hat{\sigma}_e = 0.076$)

$$t_a = \frac{0.764\sqrt{[20(3.354)]}}{0.076\sqrt{1,500.50}} = \frac{6.257}{2.944} = 2.125$$

$$t_b = \frac{0.829\sqrt{3.354}}{0.076} = 19.98$$

we can see that in the second case we reject the null hypothesis at the 1% level. If we had chosen the 5% level of significance we would have rejected H_0 for a also. This is reassuring confirmation of our simple hypothesis. We expect that if income were zero people would consume a positive quantity, and that the marginal propensity to consume should also be greater than zero. However, we should note that this is an aggregate function, and that we have no observations even remotely in the range of $Y = 0$. Thus it is not surprising that the t-statistic for b is much larger than that for a.

On p. 111 we originally suggested that $0 < b < 1$. We have tested the hypothesis that $b = 0$, so let us now test the hypothesis that $b \geqslant 1$ to complete the analysis. In this case our critical value for t is -2.55, and our test statistic

$$t_b = \frac{-0.171\sqrt{3.354}}{0.076} = -4.12$$

So again we reject the null hypothesis that the marginal propensity to consume is unity or larger at the 1% level of significance.

6.6 Prediction

Up to this point we have concentrated on the establishment of the form of the relationship between two variables, but we have not considered what uses can be made of these estimates. One of the most important uses is to predict what would occur in a future or different situation. For example if personal disposable income were to rise by 5% in the next year what would be the level of consumption? The answer to this may seem immediately apparent. We have estimated that

$$C = 0.764 + 0.829\,Y$$

hence if we substitute in the required value for Y, let us say Y^*, which was 10.06 £'000 mn (1.05 times the value of Y in the 4th quarter of 1972) for the 4th quarter of 1973, we can say that the resulting value of C, $C^* = 0.764 + 0.829\,Y^* = 9.104$ £'000 mn in this case, given that the basic relationship (3) still holds. However, we have only estimated a and b. Our estimates may not be correct. We showed in the previous section how the estimates were distributed. Thus although the point estimate may be the best we can provide, it might be more appropriate to give an interval estimate to which we can ascribe a particular level of confidence.

Obviously the size of our confidence interval will depend upon the closeness of fit of our line to the observations. The larger σ_e^2 the wider the confidence interval. Our estimate C^* depends upon not only $\hat{b}$ but $\hat{a}$ as well, both of which could be wrong. Most importantly it depends upon Y^*. Normally we are trying to predict the mean value of C^* and are not trying to find a confidence interval for a particular value of C associated with Y^*, and we shall therefore begin with the first case. The variance of C^* depends on the variance of $\hat{a}$, the variance of $\hat{b}$, the covariance of $\hat{a}$ and $\hat{b}$, and the value of Y^*,

$$\text{Var}(C^*) = \text{Var}(\hat{a}) + Y^{*2}\,\text{Var}(\hat{b}) + 2Y^*\,\text{Cov}(\hat{a},\,\hat{b})$$

$$= \sigma_e^2 \left[\frac{1}{N} + \frac{(Y^* - \overline{Y})^2}{\Sigma(Y_i - \overline{Y})^2} \right] \tag{47}$$

Since σ_e^2 is unknown we must use $\hat{\sigma}_e^2$ and

$$t = \frac{\hat{C}^* - C^*}{\hat{\sigma}_e\sqrt{[1/N + (Y^* - \overline{Y})^2/\Sigma(Y_i - \overline{Y})^2]}} \tag{48}$$

is distributed as t with $N - 2$ degrees of freedom. Thus our confidence interval for C^* at the $100(1 - \alpha)\%$ level is

$$\hat{a} + \hat{b}Y^* \pm t_{\alpha/2}\hat{\sigma}_e \sqrt{\left[\frac{1}{N} + \frac{(Y^* - \overline{Y})^2}{\Sigma(Y_i - \overline{Y})^2}\right]} \tag{49}$$

Thus in our case if we choose $\alpha = 0.05$, the 95% confidence interval for C^* where $Y^* = 10.06$ is, substituting in (49),

$$0.764 + (0.829)(10.06) \pm (2.10)(0.076) \sqrt{\left[\frac{1}{20} + \frac{(10.06 - 8.652)^2}{3.354}\right]}$$

$$= 9.104 \pm 0.128$$

or

$$8.976 \leqslant C^* \leqslant 9.232$$

If we look at (49) we can see that the confidence interval has a minimum value when $Y^* = \overline{Y}$, and widens at an increasing rate as $|Y^* - \overline{Y}|$ increases. Hence if we were to draw the confidence limits for all values of C^* estimated from values of Y^*, they would have the pattern shown in Figure 6.3. Thus the further away we try to move from the experience of our sample, the more inaccurate are our predictions, even allowing for the fact that we have assumed that the relationship is unchanged.

If we are concerned with the confidence interval for a particular value of C^* for a given Y^*, then we must allow for the fact that the

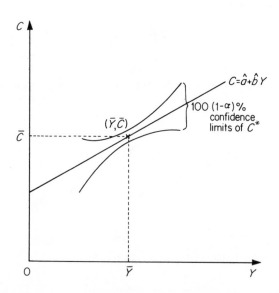

Figure 6.3

residual error exists and

$$\text{Var}(C^*) = \text{Var}(e^*) + \text{Var}(\hat{a}) + Y^{*2}\,\text{Var}(\hat{b}) + 2Y^*\text{Cov}(\hat{a}, \hat{b})$$

$$= \sigma_e^2 \left[1 + \frac{1}{N} + \frac{(Y^* - \overline{Y})^2}{\Sigma(Y_i - \overline{Y})^2} \right] \tag{50}$$

since $E[e_i^2] = \sigma_e^2$. It is perhaps a little difficult to see the difference between this and the previous case (47). In the first case we are dealing with the variance of the mean or expected value of C^* given Y^* whereas in the present case we are dealing with the variance of any one particular C^* drawn from the distribution of C^* given Y^* and not just the mean of the distribution. Let us say that in the following year Y is in fact 10.06 £'000 mn but C is 9.26 £'000 mn. We might now wish to ask the question, has C been generated by the original relationship and is $C - C^*$ accounted for by e^* or has the relationship changed. We must allow for both error in our estimates of a and b and the random error e in the relationship. Thus if we calculate

$$t = \frac{C - C^*}{\hat{\sigma}_e\sqrt{[1 + 1/N + (Y^* - \overline{Y})^2/\Sigma(Y_i - \overline{Y})^2]}} \tag{51}$$

we can test the hypothesis $H_0 : C = C^*$.

In this case we obtain

$$t = \frac{9.26 - 9.104}{0.076\sqrt{[1 + 1/20 + (10.06 - 8.652)^2/3.354]}} = 1.60$$

Using the t table we can see that $t_{0.05}(18) = 1.73$, so we would accept the null hypothesis.

6.7 Transformation of the Data and the Shape of the Relationship

We have treated prediction somewhat uncritically. The increasing inaccuracy of estimates the further the independent variable is from the mean of the sample has been emphasized, but much depends on the accuracy of our assumptions. We have assumed that

(1) the relationship is linear,
(2) $E[e_i e_j] = \sigma^2 \ (i = j)$
$$= 0(i \neq j) \quad (i, j = 1, \ldots, N)$$
$$E[e_i] = 0$$

If assumption (1) is false our predictions may be grossly inaccurate. Figure 6.4 shows the possible degree of error where $Y = a + bX$ is a good approximation to the data over the sample period, but an appalling one over a wider range of values. We might actually predict negative values for Y when they should be positive, and the actual values of Y may lie outside our confidence interval for them.

124

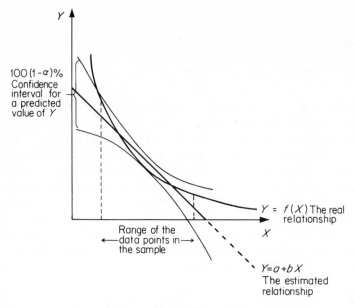

Figure 6.4

This type of situation is by no means rare; if for example we regress the output of the chemicals industry on coal consumption by the industry our estimate would suggest that if output were 50% higher coal consumption would be negative. Over the sample period a linear relationship is a good approximation to a much more complex relationship. The correlation is rather the result of the time series nature of our data than a technological relationship. Output is rising and coal consumption is an input to it, but there is not a causal link. Output is rising as a result of increased demand and coal consumption is falling as a result of technological change. The relationship we have estimated is one of association not causation. We have no hypothesis to relate changes in coal consumption to output and therefore we cannot expect to predict successfully as we have no particular reason to expect that the association will continue into the future. In fact we may very well want to argue that the trend in coal consumption will change as a result of the changing price structure between fuels. Thus we must firstly guard against confusing association over a sample period with the existence of a fundamental causal relationship.

The second pitfall is more difficult to deal with. Let us assume that we have got a specific hypothesis which appears to be borne out by the sample data. How do we know that we have specified the hypothesis in the correct form. We have assumed a linear relationship $Y = a + bX$. If we took a simple demand relationship for a good, X

$$Q_X = a + bP_X \tag{52}$$

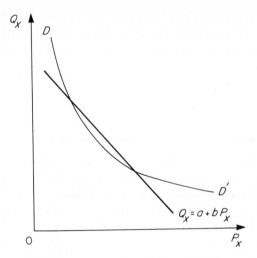

Figure 6.5   A simple demand function

where Q stands for quantity and P for price, and b is negative, although (52) might be a reasonable approximation for a narrow range of prices, we might wish to argue that it does not hold for all P_X. It would be more realistic to suggest that we have a demand curve that is convex to the origin as shown by the line DD' in Figure 6.5. The simplest form for such a curve is to assume that elasticity of demand

$$b* = \frac{dQ_X}{dP_X} \cdot \frac{P_X}{Q_X}$$

is a constant. Thus (52) should be rewritten as

$$Q_X = a*P_X^{b*} \tag{53}$$

★   This is easily shown. Differentiating (53)
★
★ $$\frac{dQ}{dP} = a*b*P^{b*-1}, \text{ then, as } \frac{P}{Q} = \frac{P}{a*P^{b*}}, \text{ we can see } \frac{dQ}{dP} \cdot \frac{P}{Q} = b*$$
★

If we take logarithms of both sides of (53)

$$\log Q_X = \log a* + b* \log P_X \tag{54}$$

we have an equation which is again in the form

$$Y = a + bX \tag{55}$$

where $Y = \log Q_X$, $a = \log a*$, $b = b*$ and $X = \log P_X$. However, (55) is a deterministic equation with no stochastic element. We must have a relationship of the form

$$Y = a + bX + v \tag{56}$$

for us to be able to use least squares, where v is a random disturbance term $N(0, \sigma^2)$. It is important to remember that although the errors are additive in the transformed model (56), they are multiplicative in the original form

$$Q_X = a * P_X^{b*} e^v \tag{57}$$

If they were additive as in (58)

$$Q_X = a * P_X^{b*} + v \tag{58}$$

we would have to use an iterative estimation method.

We have thus shown that by a simple transformation of the variables we can estimate a different shaped function. The double-logarithmic transformation is only one of several. In the last chapter (p. 91) we considered the estimation of a constant rate of growth

$$Y = ab^t e^u \tag{59}$$

If we take logarithms of both sides

$$\log Y = \log a + t \log b + u \tag{60}$$

then setting $a_1 = \log a$ and $b_1 = \log b$ we have

$$\log Y = a_1 + b_1 t + u \tag{61}$$

which is a semi-logarithmic transformation. Obviously there are others

$$Y = a + \frac{b}{X} + u \qquad \text{inverse relationship} \tag{62}$$

$$Y = a + \frac{b}{\log X} + u \quad \text{log-inverse relationship} \tag{63}$$

In each case the formulation chosen must depend upon our hypothesis. Just because the particular sample of points tends to indicate that some special shape of curve is appropriate we should not use it; we must have some hypothesis to justify the shape. If we merely wanted the best possible fit we should just use a polynomial of order $N - 1$. Clearly, on the other hand, we can use the sample as evidence to support the choice of a particular form of function. If we are estimating a demand relationship and we are uncertain whether (52) or (53) is the correct specification, the estimation of a set of $\hat{v}_i$ from (52) which were negative for low and high values of P but positive for the middle values, would incline us to the adoption of (53). On p. 91 we showed how this argument could be applied to the choice of a trend, and the reader should turn back to that page to see an illustration of the rôle of the pattern of residuals.

We can see that with all these possibilities least-squares regression is indeed a powerful tool for the estimation of economic relationships.

However, we must always be aware of the limitations of our assumptions concerning the disturbance terms. Much of the work of econometrics is concerned with estimation methods when one or other of these assumptions is violated. In Chapter 8 we shall give an outline of the way such methods proceed.

6.8 The Calculation of Least-Squares Regression

(a) By Hand and Using a Calculating Machine   We need to calculate the two regression coefficients, their standard errors, our estimate of the standard error of the e_is and lastly r^2. In fact there is little that needs to be added about their calculation at this stage. In Chapter 5 (p. 98) we considered the calculation of firstly the correlation coefficient, and secondly the covariance of the two variables and the variance of the independent variable from which we can obtain the slope of the regression line. If the equation of the line is $Y = a + bX$, there is no problem in calculating $\hat{a} = \Sigma Y/N - \hat{b}\Sigma X/N$; however, we will require two extra columns to calculate $\hat{Y}_i = \hat{a} + \hat{b}X_i$ and $\hat{e}_i = Y_i - \hat{Y}_i$ for every i in order to obtain $\hat{\sigma}_e^2 = \Sigma \hat{e}_i^2/(N-2)$. We can now proceed directly to the calculation of r^2 and the standard errors.

$$\sigma_{\hat{a}} = \sqrt{\left\{\hat{\sigma}_e^2\left[\frac{1}{N} + \frac{\overline{X}^2}{\Sigma(X_i - \overline{X})^2}\right]\right\}} \quad \text{and} \quad \sigma_{\hat{b}} = \sqrt{\frac{\hat{\sigma}_e^2}{\Sigma(X_i - \overline{X})^2}}$$

we should recall that

$$\Sigma(X_i - \overline{X})^2 = N \operatorname{Var}(X) = \Sigma X_i^2 - (\Sigma X_i)^2/N$$
$$= \Sigma(x_i - A)^2 - N(\overline{X} - A)^2$$

where A is an arbitrary origin.

(b) Using a Computer   In this case, unlike Chapter 5, we must retain the values of both variables in the computer. We shall require two loops to complete the calculations, the first to compute the regression coefficients and the second to compute their standard errors.

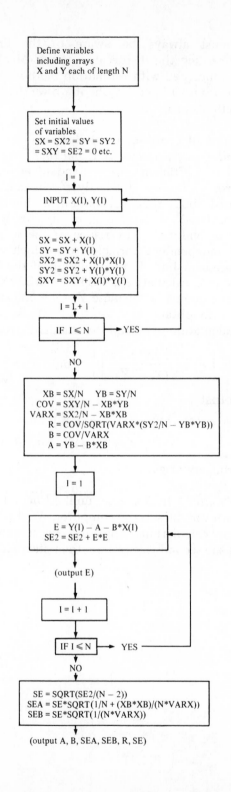

References and Suggested Reading

1. A more formal approach to the simple regression model can be found in

> J. Johnston, *Econometric Methods*, 2nd edition, McGraw—Hill, New York, 1972, Chapter 2.
>
> 2. *Digest of U.K. Energy Statistics*, H.M.S.O., quarterly.

Questions

1 Regress the output of the chemicals industry on coal consumption by the industry for the years 1956—1970. Use the regression line to suggest what the output would have been given that coal consumption was 6.72 million tons in 1955.

Output and coal consumption in the chemicals industry

Year	Coal consumption (mn tons)	Output[a]
1956	6.75	69.6
1957	6.64	72.4
1958	6.13	72.4
1959	5.97	72.9
1960	6.13	88.7
1961	5.95	90.1
1962	5.75	93.3
1963	5.44	100.0
1964	5.25	109.8
1965	5.25	117.1
1966	4.95	123.4
1967	4.62	130.4
1968	4.39	140.0
1969	3.87	148.7
1970	2.49	157.9

[a]Output is expressed as an index number, 1963 = 100.
Source: *Annual Abstract* and *Digest of U.K. Energy Statistics*

2 Suggest and then estimate the relationship between total U.K. consumers' expenditure, (E), and consumers' expenditure abroad, (A), using the information given below and in the table overleaf. Before starting the calculation specify the sort of values you expect for the regression coefficients. Test the hypotheses that, individually, the coefficients are not different from zero. Explain your results — an examination of the pattern of the residuals may prove helpful.

$\bar{A} = 328.27$, $\bar{E} = 215.36$
Var $A = 689.99$, Var $E = 251.43$
Cov$(EA) = 125.59$, $N = 11$
$\Sigma E^2 = 512{,}945$

U.K. Consumers' Expenditure 1961—71 in 1963 prices

Year	Total expenditure (E) £'00 mn	Expenditure abroad (A) £ mn
1961	189	295
1962	193	303
1963	201	326
1964	208	345
1965	212	360
1966	216	368
1967	221	326
1968	227	286
1969	228	310
1970	234	336
1971	240	356

Source: *Annual Abstract*

3   A firm with a fixed capital stock records its output during a period when the man-hours used in production vary. These figures are set out below. Use least squares to fit a Cobb—Douglas production function to the data.

(We can express the function as

$$Q = AK^a L^b \qquad (1)$$

where Q is output, K is capital stock and L is man-hours. As K is fixed we can rewrite (1) as

$$Q = cL^b \qquad (2)$$

where $c = AK^a$.)

The firm has determined that $a = 0.5$. Test the hypothesis that the firm enjoys constant returns to scale (i.e. $(a + b) = 1$).

Output (units)	Man-hours
327	54
254	35
298	47
255	37
307	49
320	51
356	60
376	65
309	52
290	44
298	46
267	38
271	39
285	43
314	49

4   Using the aggregate regional results from the 1969–70 *Family Expenditure Survey* for Great Britain, the least-squares regression of average weekly household expenditure on food on average household income per week is

$$F = 4.535 + 0.0758\,Y \tag{1}$$

where F is average weekly household expenditure on food in £ and Y is average weekly household income in £. (Sampling problems have been ignored.)

Y for Northern Ireland was £28.86. If (1) holds, give a 95% confidence interval for your estimate of F for Northern Ireland.

The actual value of F for Northern Ireland was £7.55. Test the hypothesis that the values for Northern Ireland are drawn from the same population as those for the regions of Great Britain.

$$\bar{Y} = 33.156, \quad \Sigma Y^2/N = 1{,}106.269, \quad N = 11, \quad \hat{\sigma}_e^2 = 0.04622$$

Answers

1   Let us refer to coal consumption as C and output as Q. We can set out the table we would need if we were using a calculating machine.

Year	Q	C	C^2	QC
1956	69.6	6.75	45.56	469.80
1957	72.4	6.64	44.09	480.74
1958	72.4	6.13	37.58	443.81
1959	72.9	5.97	35.64	435.21
1960	88.7	6.13	37.58	543.73
1961	90.1	5.95	35.40	536.10
1962	93.3	5.75	33.06	536.48
1963	100.0	5.44	29.59	544.00
1964	109.8	5.25	27.56	576.45
1965	117.1	5.25	27.56	614.78
1966	123.4	4.95	24.50	610.83
1967	130.4	4.62	21.34	602.45
1968	140.0	4.39	19.27	614.60
1969	148.7	3.87	14.98	575.47
1970	157.9	2.49	6.20	393.17
Σ	1,586.7	79.58	439.91	7,977.62
Σ/N	105.78	5.31	29.33	531.84

$$\hat{b} = \frac{\text{Cov}(Q, C)}{\text{Var } C} = \frac{(\Sigma QC)/N - (\Sigma Q/N)\cdot(\Sigma C/N)}{\Sigma C^2/N - (\Sigma C/N)^2}$$

$$= \frac{531.84 - (105.78)(5.31)}{29.33 - 5.31^2} = \frac{-29.32}{1.187} = -24.7$$

$$\hat{a} = \bar{Q} - \hat{b}\bar{C} = 105.78 + (24.7)(5.31) = 236.8$$

Thus the estimated line is

$$Q = 236.8 - 24.7C$$

and, using this relation,

when $C = 6.72, \quad \hat{Q} = 70.8$

2 We have the usual problems in dealing with aggregate relationships. It might be reasonable to suggest that at low levels of expenditure, expenditure abroad is zero, while at higher levels expenditure abroad increases with increases in expenditure. In aggregate we do not have to bother about low levels of expenditure, and we could perhaps suggest that the linear relationship

$$A = a + bE$$

where $a < 0$ and $0 < b < 1$, might be a fair approximation to reality. We would also expect b to be much closer to zero than unity as expenditure abroad is relatively unimportant at all levels of total expenditure.

Estimating a and b

$$\hat{b} = \frac{\text{Cov}(E, A)}{\text{Var } E} = \frac{125.59}{251.43} = 0.4995$$

$$\hat{a} = \bar{A} - \hat{b}\bar{E} = 328.27 - (0.4995)(215.36) = 220.7$$

To test if the coefficients are significantly different from zero

$$H_0: a = 0, \qquad H_0: b = 0$$
$$H_1: a \neq 0, \qquad H_1: b \neq 0$$

at say, the 5% level, we must calculate

$$t_a = \frac{(\hat{a} - a)\sqrt{[N\Sigma(E - \bar{E})^2]}}{\hat{\sigma}_e\sqrt{\Sigma E^2}} \qquad t_b = \frac{(\hat{b} - b)\sqrt{\Sigma(E - \bar{E})^2}}{\hat{\sigma}_e}$$

To calculate $\hat{\sigma}_e$ we need to compute the residuals and their squares.

	1961	1962	1963	1964	1965	1966
$\hat{A}_i$	315.1	317.1	321.1	324.6	326.6	328.6
$\hat{e}_i$	−20.1	−14.1	4.9	20.4	33.4	39.4
$\hat{e}_i^2$	404.01	198.81	24.01	416.16	1,115.56	1,552.36

	1967	1968	1969	1970	1971	
$\hat{A}_i$	333.1	334.1	334.6	337.6	340.6	
$\hat{e}_i$	−5.1	−48.1	−24.6	−1.6	15.4	
$\hat{e}_i^2$	26.01	2,313.61	605.16	2.56	237.16	

$$\Sigma\hat{e}_i = -0.1, \quad \Sigma\hat{e}_i^2 = 6,895.41, \quad \hat{\sigma}_e^2 = 6,895.41/9 = 766.16$$

$$t_a = \frac{220.7\sqrt{[(11^2)(251.43)]}}{\sqrt{766.16}\sqrt{512,945}} = 1.94 \quad t_b = \frac{0.4995\sqrt{[(11)(251.43)]}}{\sqrt{766.16}}$$

$$= 0.95$$

$t(9)_{0.025} = 2.26$ so we accept the null hypothesis in both cases.

The results are disappointing. If we look at the residuals we can see that they are firstly negative then positive, then negative and lastly positive again, and their pattern does not seem very random. The first six values are of increasing magnitude and the last six follow a U-shaped pattern. We need to think of factors which we have left out of the relationship. The U-shaped pattern should stimulate the memory — during this period a limit of £50/head was placed on expenditure abroad. Hence we would expect that these years would exhibit negative residuals. Our failure to obtain a significant linear relationship lies in the fact that the underlying relationship has been obscured by the imposition of a constraint on one of the variables. (We shall consider a method of getting round this problem in Chapter 7.) We should, however, bear in mind that A and E are measured in different units, so that, while the test of the hypothesis $H_0 : b = 0$ is correct, the value of b indicates that for every £100 mn increase in consumers' expenditure there will be a £½ mn increase in consumers' expenditure abroad, not a £50 mn increase.

3 If we take logarithms of both sides of (2)

$$\log Q = \log c + b \log L \tag{3}$$

we can fit (3) by ordinary least squares and estimate $\log c$ and b, provided that we assume that the disturbances are additive in logarithms and $N(0, \sigma_e^2)$. In this instance we have made a strong assumption.

Using the calculations shown in the table we can obtain

$$\hat{b} = \mathrm{Cov}(\log Q, \log L)/\mathrm{Var}(\log L) = \frac{328.903/15 - (5.704)(3.841)}{221.747/15 - 3.841^2}$$

$$= 0.01780/0.02985 = 0.5963$$

$$\hat{a} = \overline{\log Q} - \hat{b}\,\overline{\log L} = 5.704 - (0.5963)(3.841) = 3.414$$

Testing $H_0 : a + b = 1$ is equivalent to testing $H_0 : b = 1 - a = 0.5$. Thus our test statistic is

$$t_b = \frac{(\hat{b} - b)\sqrt{\Sigma(\log L - \overline{\log L})^2}}{\hat{\sigma}_e}$$

$$= \frac{0.0963\sqrt{[15(0.02985)]}}{0.0054} = \frac{0.0644}{0.0054} = 11.93$$

$t(13)_{0.01} = 2.65$ so we reject H_0. The firm has increasing returns to scale.

Q	L	$\log Q$	$\log L$	$(\log L)^2$	$(\log Q \cdot \log L)$	$\hat{e}$	$\hat{e}^2\ 10^{-4}$
327	54	5.790	3.989	15.912	23.096	-0.00264	0.070
254	35	5.537	3.555	12.638	19.684	0.00032	0.000
298	47	5.697	3.850	14.823	21.933	-0.01276	0.163
255	37	5.541	3.611	13.039	20.009	-0.02624	0.689
307	49	5.727	3.892	15.148	22.289	-0.00780	0.061
320	51	5.768	3.932	15.461	22.680	0.00935	0.088
356	60	5.875	4.094	16.761	24.052	0.01975	0.390
376	65	5.930	4.174	17.422	24.752	0.02704	0.731
309	52	5.733	3.951	15.610	22.651	-0.03698	1.368
290	44	5.670	3.784	14.319	21.455	-0.00040	0.000
298	46	5.697	3.829	14.661	21.814	-0.00023	0.000
267	38	5.587	3.638	13.235	20.326	0.00366	0.013
271	39	5.602	3.664	13.425	20.526	0.00316	0.010
285	43	5.653	3.761	14.145	21.261	-0.00368	0.014
314	49	5.749	3.892	15.148	22.375	0.01420	0.202
Σ		85.556	57.616	221.747	328.903		3.736
Σ/N		5.704	3.841			$\Sigma/(N-2)$	0.287

All logarithms are to the base e.

$$\hat{e} = \log Q - \widehat{\log Q}$$
$$= \log Q - \hat{a} - \hat{b} \log L$$

4 If $Y^* = 28.86$ then using (1)

$$\hat{F} = 4.535 + (0.0758)(28.86) = 6.723$$

The variance of $\hat{F}$ (the mean of F for $Y = 28.86$) is

$$\hat{\sigma}_e^2 \left[\frac{1}{N} + \frac{(Y^* - \bar{Y})^2}{\Sigma(Y - \bar{Y})^2} \right]$$

hence we can construct the 95% confidence interval for $\hat{F}$

$$\hat{F} \pm t(9)_{0.025} \sqrt{ \left\{ \hat{\sigma}_e^2 \left[\frac{1}{N} + \frac{Y^* - \bar{Y})^2}{\Sigma(Y - \bar{Y})^2} \right] \right\} }$$

In our case

$$6.723 \pm 2.26 \sqrt{ \left\{ 0.04622 \left[\frac{1}{11} + \frac{(28.86 - 33.156)^2}{11(1{,}106.269 - 33.156^2)} \right] \right\} }$$

$$= 6.723 \pm 0.280$$

or

$$6.443 \leqslant \hat{F} \leqslant 7.003$$

To test if $F = 7.55$ when $Y^* = 28.86$ is a member of our population, we must test $H_0 : F = F^*$. Our test statistic is

$$t = \frac{F - \hat{F}}{\sqrt{\left\{ \hat{\sigma}_e^2 \left[1 + \frac{1}{N} + \frac{(Y^* - \bar{Y})^2}{\Sigma(Y - \bar{Y})^2} \right] \right\}}}$$

In our case

$$t = \frac{7.55 - 6.723}{\sqrt{\left\{ 0.04622 \left[1 + \frac{1}{11} + \frac{(28.86 - 33.156)^2}{(11)(1,106.269 - 33.156^2)} \right] \right\}}} = 3.33$$

$t(9)_{0.005} = 3.25$

so we reject H_0 at the 1% level.

It is worth noting that if we test $H_0 : b = 0$, $t_b = 3.08$.

7 Multiple Regression

As the book has progressed we have moved on from the consideration of the characteristics of a single variable to the relationship between two variables. We are now in a position to consider the relationship between many variables. If a variable or variables is omitted from the relationship the assumption which we had about the disturbances that made the ordinary least squares estimators best linear unbiased are violated. We must therefore expand the relationship to include the other variables.

7.1 Multivariate Models
The bivariate relationships we considered in the last two chapters were in fact comprised of three variables, let us call them X, Y and e, in the form

$$Y = a + bX + e \tag{1}$$

where X is a known variable and Y another observable variable which depends on X in the form given by (1), but e is an unobservable random variable whose expected value is zero. We suggested that e was comprised of a large number of factors which accounted for the difference between any particular value, Y_i, and the value we would expect from the underlying relationship $Y_i^* = a + bX_i$. In some cases these factors have been included in the disturbance term, e, either because they cannot be identified properly, or because they cannot be measured properly, or lastly because the nature of their relationship with the dependent variable, Y, is unknown, and on other occasions the bivariate relationship is only used to identify the most important part of a more general relationship.

Let us take a simple example. We have suggested that there is a linear relationship between imports and gross domestic product (see p. 78).

At the time we only estimated the correlation coefficient between the two variables. The least-squares regression of imports on GDP is

$$(M = a_1 + b_1 Y + u_1)$$

$$M = -1507 + 0.4235Y, \quad r^2 = 0.921 \tag{2}$$

$$(151.5) \quad (0.0195)$$

where M is imports in constant 1963 prices, 1961 I—1972 II (quarterly),

Y is GDP in constant 1963 prices, 1961 I—1972 II (quarterly),

a_1 and b_1 are constants and u_1 is a random disturbance term independently distributed $N(0, \sigma_u^2)$.

Standard errors are shown in parentheses.

(The period of estimation is greater than that used in Chapter 5.)

Yet at a very elementary level of economics we know that price should enter into the relationship. As the price of imports rises relative to the price of domestic products we would expect the demand for imports to decrease relative to the demand for domestic products.

Thus we could reformulate (2) as

$$M = f(Y, PR) \tag{3}$$

where $PR = PM/PD$,

PM is the price of imports and

PD is the price of the domestic product.

If we assume that the relationship is linear

$$M = a_2 + b_2 Y + c_2 PR \tag{4}$$

and further, since the relationship is still stochastic,

$$M = a_2 + b_2 Y + c_2 PR + u_2 \tag{5}$$

where u_2 is $N(0, \sigma_{u_2}^2)$. If we believed that people did not react identically to rises in import prices and to falls in domestic prices we would specify the components of PR separately

$$M = a_3 + b_3 Y + c_3 Y + c_3 PM + d_3 PD + u_3 \tag{6}$$

We have changed the subscripts of the regression coefficients in (2), (5) and (6), even where the variables are unchanged, because except under special circumstances their values will be different. It is not normally the case that

$$u_1 = c_2 PR + u_2 \tag{7}$$

In fact it will only be true if Y and PR are completely uncorrelated, and a_1 will only equal a_2 if PR has a zero mean. Furthermore, unless (7) is true, if a variable is omitted from a regression equation the estimates of the coefficients of the remaining variables will be biased. Thus it is not possible to regress M on Y, then regress M on PR and hence find the regression of M on Y and PR by combining the two simple regressions. We must therefore obtain a method of estimating the multiple regression of M on Y and PR.

The method of multiple regression must be general whatever the number of explanatory variables, since in theory there is no restriction on their number. As we can see, the form of the relationship is that the left-hand, endogenous, variable is determined by the relationship with the right-hand, exogenous, variables which are predetermined outside the model. The model is multivariate in that there are several variables in a single equation, not that there is more than one bivariate equation.

7.2 The Least-Squares Method of Multiple Regression

We can set out the method of least squares in multiple regression very easily, but the actual numerical process of calculation becomes extremely laborious if there are more than two exogenous variables. In fact multiple regression analysis has only been a practical possibility in all but simple cases since the invention of the electronic computer. We shall, therefore, examine the methodology in such a way that the reader could set up a computer program to perform the appropriate calculations. Further, we shall provide sufficient information for the interpretation of the output that is obtained from the many multiple regression programs that computer installations have in their possession.

In our general model the endogenous variable Y depends upon a set of exogenous variables $X_1, X_2, \ldots, X_k$ in the linear relation

$$Y = b_0 + b_1 X_1 + b_2 X_2 + \ldots + b_k X_k + e \tag{8}$$

For a given set of n observations on each of the k variables, $X_{ij} (i = 1, \ldots, n; j = 1, \ldots, k)$ we will observe a set of n values on Y_i which will vary round the mean of $b_0 + b_1 X_{i1} + \ldots + b_k X_{ik}$ according to the distribution of e_i, which we assumed in the simple case is $N(0, \sigma^2)$ for all i. There is no reason why we should alter that assumption at this stage, or the other assumptions that $E[e_i] = 0$ and $E[e_i e_j] = 0$ $(i \neq j)$. Thus our least-squares problem is to estimate the bs such that the error sum of squares $\Sigma \hat{e}_i^2$ is a minimum, where

$$\hat{e}_i = Y_i - (\hat{b}_0 + \hat{b}_1 X_{i1} + \ldots + \hat{b}_k X_{ik}) \tag{9}$$

Again as in the two-variable case we are distinguishing between e_i, the disturbance in the real relationship, and $\hat{e}_i$, the estimated error or residual in our calculated relationship.

As in the bivariate case in the last chapter, in the process of minimization we derive as many equations in the coefficients as there are unknowns. The equations are called (see p. 115) the Normal

Equations, and whereas there are 2 of them in the bivariate case there are $k + 1$ in the multivariate case.

★ To minimize
★
★ $$\Sigma \hat{e}_i^2 = \Sigma (Y_i - \hat{b}_0 - \hat{b}_1 X_{i1} - \ldots - \hat{b}_k X_{ik})^2 \qquad (10)$$
★
★ we must differentiate (10) with respect to each $\hat{b}$ in turn, and set
★ the derivative equal to zero to find a stationary value. The
★ differentiation is straightforward
★
★ $$\frac{\partial \Sigma \hat{e}_i^2}{\partial \hat{b}_0} = 2\Sigma (Y_i - \hat{b}_0 - \hat{b}_1 X_{i1} - \ldots - \hat{b}_k X_{ik})(-1) \qquad (11)$$
★
★ $$\frac{\partial \Sigma \hat{e}_i^2}{\partial \hat{b}_1} = 2\Sigma [(Y_i - \hat{b}_0 - \hat{b}_1 X_{i1} - \ldots - \hat{b}_k X_{ik})(-X_{i1})] \qquad (12)$$
★
★ etc. Hence the Normal Equations can be formed by setting each
★ derivative equal to zero, dividing both sides by two and rearranging.

The Normal Equations will now run

$$
\begin{aligned}
N b_0 + \hat{b}_1 \Sigma X_{i1} + \ldots + \hat{b}_k \Sigma X_{ik} &= \Sigma Y_i \\
\hat{b}_0 \Sigma X_{i1} + \hat{b}_1 \Sigma X_{i1}^2 + \ldots + \hat{b}_k \Sigma X_{ik} X_{i1} &= \Sigma Y_i X_{i1} \\
\hat{b}_0 \Sigma X_{i2} + \hat{b}_1 \Sigma X_{i1} X_{i2} + \ldots + \hat{b}_k \Sigma X_{ik} X_{i2} &= \Sigma Y_i X_{i2} \\
\vdots \qquad \vdots \qquad\quad \vdots \qquad\quad \vdots \\
\hat{b}_0 \Sigma X_{ik} + \hat{b}_1 \Sigma X_{i1} X_{ik} + \ldots + \hat{b}_k \Sigma X_{ik}^2 &= \Sigma Y_i X_{ik}
\end{aligned}
\qquad (13)
$$

It is the process of solution of these equations which requires a computer if k is greater than 2 or at most 3.

A whole system of algebra exists to facilitate the examination of problems concerning systems of equations like that of (13). It is known as linear or *matrix algebra*. Many first-year students of economics will have attended a course in matrix algebra, but for those who have not we have set out a very brief explanation of its basic principles in Appendix 2.

However, we have already looked at the problem of solving (13) when $k = 1$, since this is the simple regression case. The procedure we followed was to multiply one equation by a constant and add it to the other in such a way that the result is to eliminate $\hat{b}_0$ and express $\hat{b}_1$ in terms of X and Y. Thus we have solved the system for $\hat{b}_1$. We merely need to substitute back into the other equation for $\hat{b}_1$ and we have $\hat{b}_0$ in terms of X and Y, and we have solved the system completely. (The reader should turn to p. 115 to check this.)

★ To solve three simultaneous equations, we have to multiply the first
★ equation by a constant such that when we add the first two

★ equations together we eliminate $\hat{b}_0$, and then choose a second
★ constant to multiply the first equation in such a way that when we
★ add the first and third equations together we again eliminate $\hat{b}_0$.
★ We thus have two equations in two unknowns $\hat{b}_1$ and $\hat{b}_2$, and are
★ back in our simple case and can proceed as before. Only of course
★ we have to substitute back into the first equation for both $\hat{b}_1$ and
★ $\hat{b}_2$ in order to solve for $\hat{b}_0$.

Clearly we can use this method of elimination and substitution for any number of equations, and we have set out this method, called Gauss—Jordon elimination, in Appendix 2.

It is inordinately complex to show the solution of (13) for any k greater than 2. When $k = 2$ we obtain the following expressions for the three coefficients b_0, b_1 and b_2 in the equation

$$Y_i = b_0 + b_1 X_{i1} + b_2 X_{i2} + e_i \tag{14}$$

$$\hat{b}_0 = \bar{Y} - \hat{b}_1 \bar{X}_1 - \hat{b}_2 \bar{X}_2 \tag{15}$$

$$\hat{b}_1 = \frac{A - B}{\Sigma(X_{i2} - \bar{X}_2)^2 \Sigma(X_{i1} - \bar{X}_1)^2 - [\Sigma(X_{i1} - \bar{X}_1)(X_{i2} - \bar{X}_2)]^2}$$

where

$A = \Sigma(X_{i1} - \bar{X}_1)(Y_i - \bar{Y})\Sigma(X_{i2} - \bar{X}_2)^2$
$B = \Sigma(X_{i2} - \bar{X}_2)(Y_i - \bar{Y})\Sigma(X_{i1} - \bar{X}_1)(X_{i2} - \bar{X}_2)$

$$\hat{b}_1 = \frac{\mathrm{Cov}(X_1, Y)\mathrm{Var}\,X_2 - \mathrm{Cov}(X_2, Y)\mathrm{Cov}(X_1, X_2)}{\mathrm{Var}\,X_2\,\mathrm{Var}\,X_1 - [\mathrm{Cov}(X_1, X_2)]^2} \tag{16}$$

$$\hat{b}_2 = \frac{C - D}{\Sigma(X_{i2} - \bar{X}_2)^2 \Sigma(X_{i1} - \bar{X}_1)^2 - [\Sigma(X_{i1} - \bar{X}_1)(X_{i2} - \bar{X}_2)]^2}$$

where

$C = \Sigma(X_{i2} - \bar{X}_2)(Y_i - \bar{Y})\Sigma(X_{i1} - \bar{X}_1)^2$
$D = \Sigma(X_{i1} - \bar{X}_1)(Y_i - \bar{Y})\Sigma(X_{i1} - \bar{X}_1)(X_{i2} - \bar{X}_2)$

$$\hat{b}_2 = \frac{\mathrm{Cov}(X_2, Y)\mathrm{Var}\,X_1 - \mathrm{Cov}(X_1, Y)\mathrm{Cov}(X_1, X_2)}{\mathrm{Var}\,X_2\,\mathrm{Var}\,X_1 - [\mathrm{Cov}(X_1, X_2)]^2} \tag{17}$$

where the variances and covariances are sample values.

Unfortunately these are not very elegant expressions. It is all the more unfortunate since in matrix algebra the general expression for any number of equations is so very neat.

★ If we add a new variable X_0 which has the value unity for all i we
★ can rewrite (8) as

★
★ $$Y_i = b_0 X_{i0} + b_1 X_{i1} + b_2 X_{i2} + \ldots + b_k X_{ik} + e_i \tag{18}$$
★
★ We therefore express this in matrix notation as
★
★ $$y = Xb + e \tag{19}$$

★ The Normal Equations become

★ $$X'X\hat{b} = X'y \qquad (20)$$

★ and

★ $$\hat{b} = [X'X]^{-1}X'y \qquad (21)$$

Let us illustrate the case of $k = 2$ by using the demand for imports which we specified on p. 137. The results of the simple regression of imports on GDP were shown in equation (2). Adding the third variable of relative prices as in (5) we obtain

$$M = -850.1 + 0.4063Y - 535.6PR \qquad (22)$$

where

$\Sigma M = 81{,}792$	$\Sigma Y = 340{,}360$	$\Sigma PR = 45.170$
$\Sigma M^2 = (14{,}904)(10^4)$	$\Sigma Y^2 = (25{,}384)(10^5)$	$\Sigma PR^2 = 44.419$
$\Sigma MY = (61{,}289)(10^4)$	$\Sigma MPR = 80{,}043$	$\Sigma YPR = 333{,}610$

7.3 Testing Hypotheses about the Value of Coefficients in the Multivariate Model

Equation (22) on its own does not tell us how much more we know about M. It may be that although PR has a non-zero coefficient the standard error of that coefficient is large. We might therefore accept the null hypothesis that M is not related to PR under the appropriate test. Secondly, it might be that M is related to PR in a very similar fashion to he way in which it is related to Y. Thus we would not find out anything new about the behaviour of M. We must, therefore, calculate some further statistics about the regression.

Before we do this, we should note that the properties of the least-squares estimators are maintained in the case of multiple regression. The $\hat{b}_i$ are still Best Linear Unbiased. It is a straightforward matter to prove the unbiasedness, and the linearity is immediately apparent. The 'best' characteristic of minimum variance is known as the Gauss–Markov Theorem, and we have proved this theorem in the Appendix.

★ $\hat{b} = [X'X]^{-1}X'y$ is obviously linear. Taking expectations

★
★ $$E[\hat{b}] = E([X'X]^{-1}X'[Xb + e]) \qquad (23)$$
★
★ $$= [X'X]^{-1}X'Xb + [X'X]^{-1}X'E[e]$$
★
★ $$= b + [X'X]^{-1}X'E[e]$$
★
★ $$= b$$
★

★ as $[X'X]^{-1}X'X = I$ from the definition of an inverse and $E[e] = 0$
★ Thus $\hat{b}$ is an unbiased estimator of b.

We now know that the $\hat{b}_i$ are each distributed round a mean of b_i with minimum variance among linear estimators. In the case where

$k = 2$ we can write the variance of $\hat{b}_1$ and $\hat{b}_2$ as

$$\text{Var } \hat{b}_1 = \frac{\sigma^2}{\Sigma(X_{i2} - \bar{X}_2)^2 - \Sigma(X_{i1} - \bar{X}_1)(X_{i2} - \bar{X}_2)/\Sigma(X_{i1} - \bar{X}_1)^2}$$

(24)

$$\text{Var } \hat{b}_2 = \frac{\sigma^2}{\Sigma(X_{i1} - \bar{X}_1)^2 - \Sigma(X_{i1} - \bar{X}_1)(X_{i2} - \bar{X}_2)/\Sigma(X_{i2} - \bar{X}_2)^2}$$

(25)

★
★
★
★
★
★
★
★
★
★
★
★
★
★

In the general case the variance of **b** can be derived as follows from (23)

$$\hat{b} - b = [X'X]^{-1} X'e$$

Hence the expected value of the variance—covariance matrix

$$E[(\hat{b} - b)(\hat{b} - b)'] = E([X'X]^{-1} X'ee'X[X'X]^{-1})$$

as $[X'X]^{-1}$ is symmetric

$$= [X'X]^{-1} X'E[ee'] X[X'X]^{-1}$$
$$= [X'X]^{-1} X'(\sigma^2 I)X[X'X]^{-1}$$
$$= \sigma^2 [X'X]^{-1} X'X[X'X]^{-1}$$
$$= \sigma^2 [X'X]^{-1}$$

(26)

The variance of the $\hat{b}_i$ alone does not give us sufficient information to test hypotheses concerning the values of b_i. We also need to know what their distribution is. It follows from the normal distribution of our errors that the Y_i are normally distributed and hence as the $\hat{b}_i$ are linear functions of the Y_i they are also normally distributed. We can therefore form standardized normal variates as we can with simple regression

$$Z_{\hat{b}_i} = \frac{(\hat{b}_i - b_i)}{\sigma_{\hat{b}_i}}$$

(27)

However, as before, $\sigma_{\hat{b}_i}$ depends upon knowledge of σ^2 which is usually unknown. We must therefore use the errors from our estimated equation to estimate σ^2. Again we can obtain an unbiased estimator of the form

$$\hat{\sigma}^2 = \frac{\Sigma \hat{e}_i^2}{n - k - 1}$$

(28)

where the denominator takes account of the degrees of freedom. In the simple case k had the value unity.

Using this information we can derive a t statistic

$$t_{\hat{b}_i} = \frac{(\hat{b}_i - b_i)}{\hat{\sigma}_{\hat{b}_i}} \quad \text{with } n - k - 1 \text{ degrees of freedom}$$

Thus in our import example

$$\hat{\sigma}_{\hat{b}_1} = 0.0230$$

$$\hat{\sigma}_{\hat{b}_2} = 391.0$$

where $\hat{\sigma} = 82.49$. If we test the hypotheses $H_0 : b_1 = 0$ and $H_0 : b_2 = 0$ we obtain

$$t_{\hat{b}_1} = 17.68 \quad \text{and} \quad t_{\hat{b}_2} = 1.37$$

If we want the significance level to be 5% the critical value for t with $n - k - 1 = 43$ degrees of freedom is 2.02. We therefore reject the null hypothesis that $b_1 = 0$ but accept the hypothesis that $b_2 = 0$.

7.4 Joint Hypotheses

We have thus answered the first of our two questions concerning the effect of PR upon M. We have tested the hypothesis that the regression coefficient is zero. However, we have not considered a joint hypothesis concerning both $\hat{b}_1$ and $\hat{b}_2$. In Chapter 5 (p. 85) we derived a test of the hypothesis that there is no linear relationship between two variables, i.e. $H_0 : r = 0$, where r is the correlation coefficient. This test is exactly equivalent to testing the hypothesis $H_0 : b = 0$ where b is the slope of the simple regression equation.

★ The two test statistics are

$$t_r = \frac{r - 0}{\sqrt{[(1 - r^2)/(n - 2)]}} \tag{29}$$

and

$$t_b = \frac{\hat{b} - 0}{\sqrt{[\hat{\sigma}^2 / \Sigma (X_i - \bar{X})^2]}} \tag{30}$$

where

$$r = \frac{\hat{b}\sqrt{\Sigma (X_i - \bar{X})^2}}{\sqrt{\Sigma (Y_i - \bar{Y})^2}} \cdot = \frac{\sqrt{[\Sigma (Y_i - \bar{Y})^2 - \Sigma \hat{e}_i^2]}}{\sqrt{\Sigma (Y_i - \bar{Y})^2}} \quad \text{and} \quad \hat{\sigma}^2 = \frac{\Sigma \hat{e}_i^2}{n - 2}$$

Multiplying both numerator and denominator in (30) by $\sqrt{\Sigma (X_i - \bar{X})^2} / \sqrt{\Sigma (Y_i - \bar{Y})^2}$ we obtain

$$\frac{r - 0}{\sqrt{\hat{\sigma}^2} / \sqrt{\Sigma (Y_i - \bar{Y})^2}} = \frac{r - 0}{\sqrt{[\Sigma \hat{e}_i^2 / \Sigma (Y_i - \bar{Y})^2]} / \sqrt{(n - 2)}}$$

$$= \frac{r - 0}{\sqrt{[(1 - r^2)/(n - 2)]}} \tag{31}$$

In a similar fashion we can relate the hypothesis $\hat{b}_1 = \hat{b}_2 = \ldots = \hat{b}_k = 0$ to one concerning R^2 the coefficient of multiple determination where

$$R^2 = \frac{\Sigma(Y_i - \bar{Y})^2 - \Sigma\hat{e}_i^2}{\Sigma(Y_i - \bar{Y})^2} \tag{32}$$

In effect R^2 is the ratio of explained variation in Y to total variation since $\Sigma\hat{e}_i^2$ is the residual, unexplained, variation. As a result R^2 is widely used as an indicator of the success of the regression in explaining the variation in the endogenous variable. The appropriate test statistic is

$$F = \frac{R^2/k}{(1 - R^2)/(n - k - 1)} \tag{33}$$

We have not come across the distribution of F before, although it is related to both the normal and t distributions. The distribution, which is the ratio of two variates, is governed by two sets of degrees of freedom, one for each of the variates. The F distribution described above has k and $n - k - 1$ degrees of freedom respectively, the first referring to the numerator and the second to the denominator of the ratio. A table of the F distribution can be found at the end of book as Appendix Table A3.4. F can hold any value between zero and infinity, but its probability distribution is concentrated at the lower end of the range. Thus if we choose a critical value according to our required significance level we will accept the null hypothesis if our value of F lies below this value, and reject it if F is equal to, or greater than, the critical value.

This can be shown more clearly from our import demand model (22)

$$R^2 = 0.925$$

and hence

$$F = \frac{R^2/k}{(1 - R^2)/(n - k - 1)} = 265$$

Let us test the null hypothesis that $\hat{b}_1 = \hat{b}_2 = 0$ at the 1% level of significance. We can see from Appendix Table A3.4 that the appropriate value of F for 2 and 43 degrees of freedom is 5.15. Hence as $265 > 5.15$ we reject the null hypothesis.

7.5 The Output from a Multiple Regression Computer Program

This completes the information we require about our multiple regression equation if all the assumptions behind the model are true (we shall consider statistics which would indicate a departure from the assumptions in the next chapter). A typical computer output might therefore have the following form for our example

DEPENDENT VARIABLE IS M

INDEPENDENT VARIABLE	ESTIMATED COEFFICIENT	STANDARD ERROR	T-STATISTIC
CONSTANT	−850.1	502.6	−1.691
PR	−535.6	391.0	−1.370
Y	0.4063	0.0230	17.676

R SQUARED = 0.925
F−STATISTIC (2, 43) = 265
NUMBER OF OBSERVATIONS = 46
STANDARD ERROR OF REGRESSION = 82.5

7.6 Dummy Variables

Up to this point most of the variables we have considered are measurable quantities. The trend variable t on the other hand was merely constructed in such a way as to obtain a linear trend. Thus the particular values do not matter as long as they are used consistently, the slope of the trend will be the same whether we use 1, 2, 3 etc or 1961, 1962, 1963 and so on, although of course the constant term will be affected. There are other 'constructed' variables of this sort of form that we can use to take account of particular influences in our models.

Our model of import demand encompasses the devaluation of 1967. It quickly became apparent to economists that previous generalizations about importers' behaviour did not explain the changes stemming from the devaluation at all well. Our simple model substantially under-predicts the level of imports in the period after devaluation. It is clear that the parameters of the model have changed as a result of devaluation. This is not to say that the economic relationship between income, prices and imports no longer holds, but that some or all of the coefficients may have changed their value.

The simplest form that such a change could take would be a uniform increase in demand at all levels of the other explanatory variables. Thus, whereas our original model (5) ran

$$M = b_0 + b_1 Y + b_2 PR + e \qquad (34)$$

in the post devaluation situation we have

$$M = b_0 + b_1 Y + b_2 PR + b_3 + e \qquad (35)$$

where $b_3 > 0$. We have to construct some variable, say D, such that we can account for both situation (34) and situation (35) by the same equation, and hence estimate the model by our standard least-squares procedure. If D holds the value unity when (35) applies and zero when (34) applies we can see that these conditions are met. A variable such as D is known as a dummy variable.

If we re-estimate (5) in the form

$$M = b_0 + b_1 Y + b_2 PR + b_3 D + e \qquad (36)$$

we can see immediately that the standard errors of the coefficients are improved, the values of the coefficients correspond more closely to our theory, and R^2 is increased.

$$M = 1,440 + 0.2258Y - 1,565PR + 230.7D, \ R^2 = 0.965$$
$$(488) \quad (0.0313) \quad (311) \quad (34.5)$$

where the numbers in brackets are standard errors (this procedure is customary, but the reader should beware of occasions where authors place the t statistics in that position). We would now reject the null hypothesis of equality to zero for all four coefficients at the 1% level of significance.

We could also use dummy variables if we think that the coefficient of one of the other explanatory variables has changed. For example if we believe that the income coefficient, b_1, increased after devaluation we could estimate the equation

$$M = b_0 + b_1 Y + b_2 PR + b_3 D + b_4 DY + e \qquad (37)$$

where $b_4 > 0$. When $D = 0$ the income coefficient is b_1 and when $D = 1$ it is $b_1 + b_4$.

7.7 Categorical Data

A second major area of the use of dummy variables is in the analysis of categorical data. A major area of such data lies in the answers to questionnaires. Many questions are phrased in such a way that there can be only two possible answers. The question 'Do you earn more than £2,000 per year?' can only be answered by 'Yes' or 'No'. Similarly the request to state one's sex can only be answered by 'Male' or 'Female'. These answers are mutually exclusive, a person cannot be both male and female. We can, therefore, represent these two possible replies in binary form, for example, 1 for 'yes' and 0 for 'no', or 1 for 'male' and 0 for 'female'.

We can thus incorporate into our model a constant effect which is associated with one of the answers to our question but not the other. Let us say for example, that we think that for given levels of income per head and relative prices married people spend less on clothes than unmarried people. Thus

$$M = b_0 + b_1 Y + b_2 P + b_3 M + e \qquad (38)$$

where C is expenditure on clothes (per head, constant prices),

   Y is disposable income (per head, constant prices),
   P is the ratio of the price of clothing to the retail price index,
   M is a dummy variable which has the value 1 if a person is married and 0 if they are not.

If a person is married then his or her expenditure is lower by b_3 (as $M = 1$) than it would be if he or she were unmarried ($b_3 M = 0$ as $M = 0$). Thus by testing the null hypothesis that $b_3 \geqslant 0$ we can see if our assertion is borne out by our observations.

The binary form of the variables that we have used thus far is not the only possible choice. In many questionnaires the respondent is asked to grade his response by the box below which most closely corresponds to his attitude towards a given statement.

strongly disagree	disagree	neither agree nor disagree	agree	strongly agree

We can grade our dummy variable in the same way, for example by giving the responses, in order from the left, the values $-2, -1, 0, 1, 2$. This scale is arbitrary, we have chosen to make it linear.

It is also possible to use any number of categorical exogenous variables. However, we may be interested in responses to combinations of questions, sometimes known as the interaction effect between variables. Let us say that we think that salaries of economists of a particular age group are dependent upon whether the employee has a degree and the sex of the employee. Specifically we expect that graduates are paid more than non-graduates and men are paid more than women. There are four possible combinations of the two variables. In addition to seeing the effect of being a graduate or being male separately on salary, we want to determine for example what the additional effect is of being male given that one is a graduate. We must, therefore, construct four variables to allow for all the possibilities. So our model runs

$$S_i = b_1 X_{i1} + b_2 X_{i2} + b_3 X_{i3} + b_4 X_{i4} + e_i \qquad (39)$$

where S stands for salary, and i refers to the employees,

X_{i1} has the value 1 for all i,

X_{i2} has the value 1 when the employee is a graduate and 0 otherwise,

X_{i3} has the value 1 when the employee is male and 0 otherwise,

X_{i4} has the value 1 when the employee is both a graduate and male and 0 otherwise.

Thus b_1 is the salary that a non-graduate female is estimated to receive. The effect of being a graduate is b_2 and that of being male b_3. However given that one is a graduate the effect of being male is $b_3 + b_4$ and given that one is male the effect of being a graduate is $b_2 + b_4$.

We have only considered exogenous dummy variables, and not endogenous ones. The reason is very simple. You will recall that in our model, $Y = a + bX$, $Y \mid X$ is distributed as $N(a + bX, \ \sigma^2)$. However, if Y

can only have the values 0 or 1 it cannot be distributed normally for any given X. This does not mean that we cannot use regression, since $E[\hat{b}]$ is still b, and $E[\hat{a}] = a$ but the sampling distributions of $\hat{a}$ and $\hat{b}$ are unknown, and hence we cannot test hypotheses concerning them.

7.8 Seasonal Variation

A further common use of dummy variables is to take account of seasonal variation. If we believe that time series observations on a variable or variables in our multivariate linear model fluctuate consistently with the time of year in which they are observed then we should take account of this in our regression equation. We can do this in two ways, firstly by subtracting the seasonal fluctuations from each variable before running the regression or secondly by including a set of dummy variables in the regression to 'explain' the seasonal fluctuation. If we do not allow for seasonal variation, then we are violating the assumptions concerning our error term. If the relationship varies with each season then $E[e_i] \neq 0$. Furthermore $E[e_i e_j] \neq 0$ $(i \neq j)$, since there is a relationship between the errors in consecutive seasons and therefore their expected covariance is non-zero.

The structure of our model using, say, quarterly data unadjusted for seasonal variation, is such that if we take data for each quarter separately we can run four regressions of the form

$$Y_{ij} = b_{j0} + b_1 X_{ij1} + b_2 X_{ij2} + \ldots + b_k X_{ijk} \tag{40}$$

where i refers to the year of observation and j to the quarter of the observation, the third subscript of X, $1, 2, \ldots, k$, refers to each of the respective k exogenous variables. A separate regression is run on the data for each of the quarters. The only difference between the regressions is that b_0, the constant term, is different in each quarter. Of course, we would obtain four different estimates of the other bs, one from each regression, purely as a result of sampling variation. However, since the expected value of those estimators is identical in each equation we can pool our data and run a single regression providing we allow for the different b_0s. We can express each b_0 in terms of the others, for example expressing the others in terms of the constant for the fourth quarter, b_{40}, we obtain $b_{10} = b_{40} + c_1$, $b_{20} = b_{40} + c_2$ and $b_{30} = b_{40} + c_3$. Thus if we run the regression

$$Y_{ij} = b_{04} + c_1 Q_1 + c_2 Q_2 + c_3 Q_3 + b_1 X_{ij1} + \ldots + b_k X_{ijk} \tag{41}$$

where Q_1 has the value 1 when $j = 1$ and 0 otherwise,
  Q_2 has the value 1 when $j = 2$ and 0 otherwise,
  Q_3 has the value 1 when $j = 3$ and 0 otherwise,

we have overcome the problem. Clearly the choice we make about which quarter to use as the base from which to measure the other quarterly shifts is arbitrary. However, the choice will neither affect the relative values of the seasonal coefficients nor the absolute values of the other coefficients.

In fact we already have an illustration of the use of seasonal dummies in a regression equation. You will recall that our import demand example used quarterly data, but we did not disclose then that in fact rather than use seasonally adjusted data all the regressions were run with unadjusted data and dummy variables were included for the first three-quarters of each year. Taking equation (37) for example, we should add three more coefficients to the equation, whose values are

$$+ 253.8Q_1 + 161.0Q_2 + 165.8Q_3$$
$$\quad (36.4) \quad\quad (34.8) \quad\quad (35.9)$$

R^2 is unchanged as it was calculated with the inclusion of these three variables originally, however the F statistic now has the value 98.7, which is still significant at 1%. If we calculate the appropriate t statistics by dividing the coefficients for each of the seasonal constants by their respective standard errors, we can see that all three seasonal constants are significantly different from zero at the 5% level. The fact that all three coefficients are positive merely indicates that imports are lowest in the fourth quarter of the year. (There may also be some interrelationship with the general upward trend in the dependent variable.)

7.9 Seasonal Adjustment

In some cases we will not be interested in estimating regression equations with seasonally unadjusted data, but merely concerned with estimating the seasonal variation in a particular variable. Clearly our method of adjustment will depend upon the remaining fluctuations in the variables. If we believe that a variable merely fluctuates round a constant value, and that these fluctuations are composed of a seasonal constant and a random term then we can use the same method as (41) and estimate

$$X_{ij} = b_0 + c_1 Q_1 + c_2 Q_2 + c_3 Q_3 + e_{ij} \tag{42}$$

Similarly if we believe that the fluctuations are round a linear trend, t, we can estimate

$$X_{ij} = b_0 + c_1 Q_1 + c_2 Q_2 + c_3 Q_3 + b_1 t + e_{ij} \tag{43}$$

However these methods do not give us the seasonally adjusted series (the series with the seasonal component removed), nor do they account for other types of trend. The simplest way to do this is to estimate the trend either by regression, moving average or cruder methods (see Chapter 5) and estimate the average deviation of all the values for each quarter from the trend. Thus in the case of a linear trend where the estimated equation is

$$X_{ij} = \hat{a} + \hat{b}t \tag{44}$$

and

$$\hat{e}_{ij} = X_{ij} - \hat{a} - \hat{b}t \tag{45}$$

(the $\hat{e}_{ij}$ are the estimated errors in the sample of values, not the disturbances in the model), we can calculate

$$q_j = \sum_{i=1}^{N} \hat{e}_{ij}/N$$

where N is the number of years, and $X_{ij}^* = X_{ij} - q_j$ will be our seasonally adjusted series. We should add the proviso that if our method of obtaining the trend is such that

$$\sum_i \sum_j \hat{e}_{ij} \neq 0$$

then we should add an amount k to each q such that the seasonal constants sum to zero for each year. Thus if we find after calculating

$$q_j = \sum_{i=1}^{N} \hat{e}_{ij}/N$$

that

$$\sum_j q_j = m$$

where $m \neq 0$, if we define $m = 4k$ we can calculate our seasonal constants as $q_j^* = q_j - k$ since

$$\sum_j q_j^* = \Sigma(q_j - k) = 0$$

7.10 Moving Averages and Seasonal Data

This latter situation is best exemplified by the results of the method of moving averages. In Table 7.1 we have calculated a four-quarter moving average for passenger journeys from the U.K. to the European continent and Mediterranean Sea area by sea. It is immediately obvious that there is a very strong seasonal pattern to the unadjusted data as shown in column (1). This reflects the great importance of the summer tourist traffic. If we look at the trend in column (4), whose calculation is shown in columns (2) and (3), we can see that there is also a steady upward movement in the series. Column (5) shows the difference between the unadjusted series and the trend.

We must now average these fluctuations for each quarter as is shown at the bottom of the table. However the sum of the average fluctuations over all four quarters together is −29.7. If we used these averages as seasonal constants we would have fluctuations centred round values below the trend. Centring them on the trend we add $29.7/4 = 7.4$ to

Table 7.1
United Kingdom passenger movement by sea to the European continent and Mediterranean Sea area (thousands)

Time	Unadjusted	Σ4	Σ8	Moving average	Total fluctuation	Seasonal fluctuation	Seasonally adjusted	Residual fluctuation
	(1)	(2)	(3)	(4)	(5) = (1) − (4)	(6)	(7) = (1) − (6)	(8) = (5) − (6) = (7) − (4)
1968 I	262					−714	976	
II	958	3,553				127	831	
III	1,978	3,617	7,170	896.25	1,081.75	1,246	732	−164.25
IV	355	3,831	7,448	931	−576	−659	1,014	83
1969 I	326	4,135	7,966	995.75	−669.75	−714	1,040	44.25
II	1,172	4,197	8,332	1,041.5	130.5	127	1,045	3.5
III	2,282	4,348	8,545	1,068.125	1,213.875	1,246	1,036	−32.125
IV	417	4,413	8,761	1,095.125	−678.125	−659	1,076	−19.125
1970 I	477	4,677	9,090	1,136.25	−659.25	−714	1,191	54.75
II	1,237	4,758	9,435	1,179.375	57.625	127	1,110	−69.375
III	2,546	4,696	9,454	1,181.75	1,364.25	1,246	1,300	118.25
IV	498	4,896	9,592	1,199	−701	−659	1,157	−42
1971 I	415	4,891	9,787	1,223.375	−808.375	−714	1,129	−94.375
II	1,437	4,935	9,826	1,228.25	208.75	127	1,310	81.75
III	2,541	5,034	9,969	1,246.125	1,294.875	1,246	1,295	48.875
IV	542	4,969	10,003	1,250.375	−708.375	−659	1,201	−49.375
1972 I	514	5,143	10,112	1,264	−750	−714	1,228	−36
II	1,372	5,189	10,332	1,291.5	80.5	127	1,245	−46.5
III	2,715					1,246	1,469	
IV	588					−659	1,247	

Column (5) rearranged by quarter

	I	II	III	IV	TOTAL
1968			1,081.75	−576	
1969	−669.75	130.5	1,213.875	−678.125	
1970	−659.25	57.625	1,364.25	−701	
1971	−808.375	208.75	1,294.875	−708.375	
1972	−750	80.5			
Total	−2,887.4	477.4	4,954.8	−2,663.5	
Average fluctuation	−721.8	119.3	1,238.7	−665.9	−29.7
+ Adjustment	7.4	7.4	7.4	7.4	0.0
Seasonal constants	−714.4	126.7	1,246.1	−658.5	−0.1

Source: *Monthly Digest*

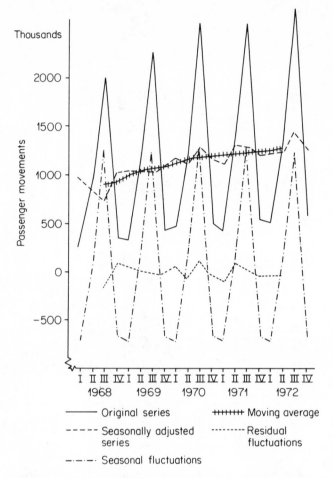

Figure 7.1   United Kingdom passenger movement by sea
to European continent and Mediterranean Sea area.
Source: *Monthly Digest*

each giving us the seasonal constants, which are reproduced for each
observation in column (6). Subtracting the seasonal constants we obtain
the seasonally adjusted series in column (7). We should note that both
the unadjusted and adjusted series still contain the residual element as
well as the trend. The effects of the seasonal adjustment can be seen in
Figure 7.1 where both series have been graphed. In addition we have
shown the trend, the seasonal constants and the residuals so that all the
components of the original series can be appreciated. It is arguable that
the seasonal constants should not be additive as we have a rising trend,
but multiplicative. In this case, as with moving averages, we should
therefore take logarithms of the original series before performing the
calculations shown in Table 7.1.

7.11 Calculating Techniques

(a) By Hand   The amount of arithmetic required is now so great as to rule out calculation by hand except in special circumstances.

(b) Calculating Machine   Again it is only reasonable to tackle the three-variable case by using this technique.

Where our model is

$$Y = b_0 + b_1 X_1 + b_2 X_2 + e$$

we wish to calculate $\hat{b}_1$, $\hat{b}_2$, $\hat{b}_0$, $\hat{e}_i$, $\hat{\sigma}^2$, $\hat{\sigma}_{\hat{b}_1}$, $\hat{\sigma}_{\hat{b}_2}$, $t_{\hat{b}_1}$, $t_{\hat{b}_2}$, R^2, F.

$$\hat{b}_1 = \frac{\text{Cov}(X_1, Y)\text{Var } X_2 - \text{Cov}(X_2, Y)\text{Cov}(X_1, X_2)}{\text{Var } X_1 \text{ Var } X_2 - [\text{Cov}(X_1, X_2)]^2}$$

$$\hat{b}_2 = \frac{\text{Cov}(X_2, Y)\text{Var } X_1 - \text{Cov}(X_1, Y)\text{Cov}(X_1, X_2)}{\text{Var } X_1 \text{ Var } X_2 - [\text{Cov}(X_1, X_2)]^2} \tag{46}$$

We must therefore set up a table of columns of $Y, X_1, X_2, Y^2, X_1^2, X_2^2$, $X_1 Y, X_2 Y$ and $X_1 X_2$ and calculate their sums. This will give us

$$\text{Cov}(X_1, X_2) = \frac{\Sigma X_1 X_2}{N} - \frac{\Sigma X_1}{N} \frac{\Sigma X_2}{N}$$

and so on (see Chapter 5, p. 99), where N is the number of observations that we have on the variables. Hence we can compute the elements of $\hat{b}_1$ and $\hat{b}_2$.

$$\hat{b}_0 = \bar{Y} - \hat{b}_1 \bar{X}_1 - \hat{b}_2 \bar{X}_2 \tag{47}$$

$$\hat{e}_i = Y_i - \hat{b}_0 - \hat{b}_1 X_{i1} - \hat{b}_2 X_{i2} \tag{48}$$

$$\hat{\sigma}^2 = \Sigma \hat{e}_i^2 / (N - 3) \tag{49}$$

$$\hat{\sigma}_{\hat{b}_1} = \frac{\hat{\sigma}}{N \text{ Var } X_2 - \text{Cov}(X_1, X_2)/\text{Var } X_1}$$

$$\hat{\sigma}_{\hat{b}_2} = \frac{\hat{\sigma}}{N \text{ Var } X_1 - \text{Cov}(X_1, X_2)/\text{Var } X_2} \tag{50}$$

$$t_{\hat{b}_1} = \hat{b}_1 / \hat{\sigma}_{\hat{b}_1} \qquad t_{\hat{b}_2} = \hat{b}_2 / \hat{\sigma}_{\hat{b}_2} \tag{51}$$

$$R^2 = \frac{N \text{ Var } Y - \Sigma \hat{e}_i^2}{N \text{ Var } Y} \tag{52}$$

$$F = \frac{R^2/2}{(1 - R^2)/(N - 3)} \tag{53}$$

(c) By Computer   Using the model $Y = Xb + e$ where there are N observations on each variable and K variables in X, the first of which is a column of ones, we shall employ a procedure which uses subprograms for the matrix operations.

154

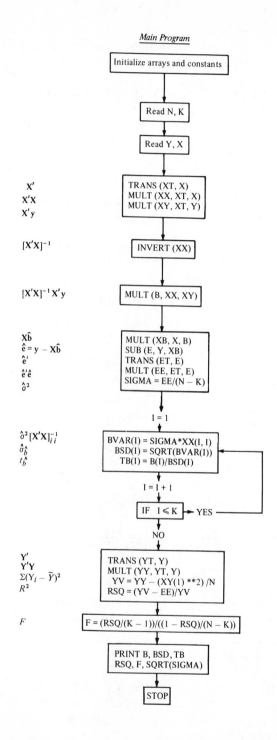

Main Program

Initialize arrays and constants

Read N, K

Read Y, X

$\mathbf{X}'$
$\mathbf{X}'\mathbf{X}$
$\mathbf{X}'y$

TRANS (XT, X)
MULT (XX, XT, X)
MULT (XY, XT, Y)

$[\mathbf{X}'\mathbf{X}]^{-1}$

INVERT (XX)

$[\mathbf{X}'\mathbf{X}]^{-1}\mathbf{X}'y$

MULT (B, XX, XY)

$\mathbf{X}\hat{\mathbf{b}}$
$\hat{e} = y - \mathbf{X}\hat{\mathbf{b}}$
$\hat{e}'$
$\hat{e}'\hat{e}$
$\hat{\sigma}^2$

MULT (XB, X, B)
SUB (E, Y, XB)
TRANS (ET, E)
MULT (EE, ET, E)
SIGMA = EE/(N − K)

I = 1

$\hat{\sigma}^2[\mathbf{X}'\mathbf{X}]_{ii}^{-1}$
$\hat{\sigma}_{\hat{b}}$
$t_{\hat{b}}$

BVAR(I) = SIGMA*XX(I, I)
BSD(I) = SQRT(BVAR(I))
TB(I) = B(I)/BSD(I)

I = I + 1

IF   I ⩽ K   → YES

NO

$\mathbf{Y}'$
$\mathbf{Y}'\mathbf{Y}$
$\Sigma(Y_i - \bar{Y})^2$
R^2

TRANS (YT, Y)
MULT (YY, YT, Y)
YV = YY − (XY(1) **2) /N
RSQ = (YV − EE)/YV

F

F = (RSQ/(K − 1))/((1 − RSQ)/(N − K))

PRINT B, BSD, TB
RSQ, F, SQRT(SIGMA)

STOP

Subprograms

(1) TRANS to form transpose of a matrix ($\mathbf{B} = \mathbf{A'}$)

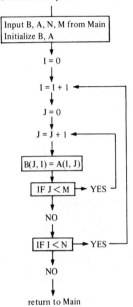

Input B, A, N, M from Main
Initialize B, A

I = 0

I = I + 1

J = 0

J = J + 1

B(J, I) = A(I, J)

IF J < M → YES

NO

IF I < N → YES

NO

return to Main

(2) MULT to multiply two matrices (**A** = **BC**)

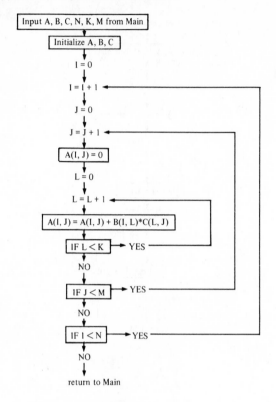

(3) SUB to subtract two matrices (**A** = **B** – **C**)

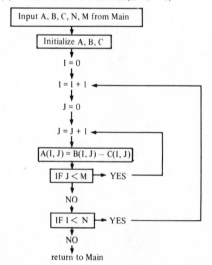

(4) INVERT to compute inverse of a matrix A

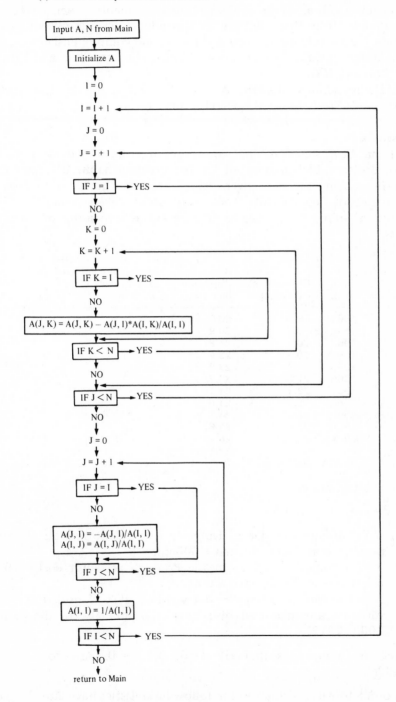

References and Suggested Reading

1. Matrix Algebra. There are many texts available which cover this subject more fully than we have in Appendix 2. Two are listed below the first is relatively elementary and the second comprehensive.

G. Mills, *Introduction to Linear Algebra*, George Allen and Unwin, London, 1969.

G. Hadley, *Linear Algebra*, Addison—Wesley, Reading, Mass., 1961.

2. *Financial Statistics*, H.M.S.O., monthly.

Questions

1. It has been suggested that consumption is not only determined by personal disposable income but also by previous habits. Examine this hypothesis using the data given below. Explain whether the values of the regression coefficients that you obtain seem plausible. Use consumers' expenditure, lagged one period as a measure of previous habits.

Year	Consumers' expenditure	Personal disposable income	Seasonally adjusted 1963 prices £'000mn
1961	18.9	20.7	
1962	19.3	20.9	
1963	20.1	21.8	
1964	20.8	22.7	
1965	21.2	23.2	
1966	21.6	23.7	
1967	22.1	24.1	
1968	22.7	24.6	
1969	22.8	24.7	
1970	23.4	25.6	
1971	24.0	26.3	

Source: *Economic Trends*

2. Using the model

$$V = a + bY + cR + e,$$

where V is investment — Gross Domestic Capital Formation at 1970 prices — seasonally adjusted £'000mn,

Y is income — Gross Domestic Product, at factor cost — 1970 prices seasonally adjusted £'000mn,

R is the rate of interest — flat yield on 2½% Consols (%) — (all variables are measured over quarterly intervals for the period 1963—1972).

Test the hypotheses that (i) $b = 0$, (ii) $c = 0$. Are your results plausible?

In order to ease calculation the following statistics have already been obtained:

$$N = 40, \qquad\qquad \Sigma(V_i - \bar{V})^2 = 2.16$$

$$\Sigma(Y_i - \bar{Y})^2 = 19.2, \qquad \Sigma(R_i - \bar{R})^2 = 76.36$$

$$\Sigma(V_i - \bar{V})(R_i - \bar{R}) = 11.0, \quad \Sigma(V_i - \bar{V})(Y_i - \bar{Y}) = 6.24$$

$$\Sigma(Y_i - \bar{Y})(R_i - \bar{R}) = 35.07, \quad \hat{\sigma}_e^2 = 0.0032$$

The original data were obtained from *Economic Trends*.

3. It is possible to suggest that the rate of change of wages, w, depends upon the level of unemployment, U, the rate of change of retail prices, p, and the rate of change of the proportion of the labour force that belongs to a trade union, N. Thus

$$w = a + bU + cp + dN + e$$

If we estimate this relationship using quarterly data for the United Kingdom from 1948 to 1967 we can obtain the following equation for periods when there was no specific government incomes policy

$$w = 6.672 - 2.372U + 0.475p + 0.136N$$
$$(1.152) \ (0.652) \quad (0.076) \quad (1.943)$$

$R^2 = 0.856$, Number of observations = 31
(Standard errors in parentheses.)

When there was an incomes policy the equation runs

$$w = 3.919 - 0.404U + 0.227p + 3.764N$$
$$(1.726) \ (0.721) \quad (0.244) \quad (2.335)$$

$R^2 = 0.138$, Number of observations = 37

(i) Do any of the individual variables affect the rate of change of wages?

(ii) Do the coefficients have signs you would expect?

(iii) Do the equations as a whole explain any variation in w (i.e. does $b = c = d = 0$)?

(iv) Do the differences between the two equations suggest whether incomes policies have worked?

(The model and equations are derived from Lipsey and Parkin 'Incomes Policy a rè-appraisal', *Economica*, May 1970.)

4. Using questionnaire replies listed below examine the linear relationship between annual earnings, education and sex. Firstly explain the effect that you would expect education and sex to have on earnings. Secondly formulate the model so that the interaction between education and sex can also be estimated. The variables are defined as

follows:

> Y = annual earnings in tax year 1972/3 of people attaining their 28th birthday during the year
> M = male
> F = female
> U = university first degree
> P = postgraduate qualification Diploma, M.A., Ph.D., etc. (all those who are P are also U)

Questionnaire responses — X indicates a positive response

Questionnaire No.	$Y(£)$	M	F	U	P	Questionnaire No.	$y(£)$	M	F	U	P
1	2,500	X			X	18	1,200		X		
2	1,400		X			19	1,200	X			
3	1,600		X	X		20	2,200		X	X	
4	2,800	X			X	21	2,800		X		X
5	2,600	X		X		22	2,400	X		X	
6	2,400		X	X		23	2,800	X		X	
7	2,000	X				24	3,000	X			X
8	2,200	X				25	1,500		X		
9	2,400		X	X		26	1,300	X			
10	1,800		X			27	1,600	X			
11	2,800	X			X	28	2,600	X		X	
12	3,200	X		X		29	1,200		X	X	
13	1,500		X			30	3,000	X		X	
14	1,700	X				31	2,800	X		X	
15	2,600	X			X	32	2,000	X			
16	2,000	X									
17	2,000		X	X							

5. The table below sets out the quarterly changes in the money supply (defined as M3) in the U.K. over the period 1968 IV–1973 III. Is there any evidence that the series needs seasonally adjusting? If so is

United Kingdom money stock (M3)
Changes during period (£mn)

Period	Changes in M3	Period	Changes in M3
1968 IV	649	1971 I	15
1969 I	−301	II	471
II	−52	III	452
III	199	IV	1,428
IV	657	1972 I	460
1970 I	−430	II	1,698
II	733	III	955
III	387	IV	2,186
IV	896	1973 I	896
		II	1,528
		III	2,345

Source: *Financial Statistics*

it reasonable to use a linear trend? Compare the use of linear and moving average trends. (A graph of the series will help.)

Lastly, can you explain the seasonal fluctuations in the money supply?

Answers

1. Let us denote consumers' expenditure by C and personal disposable income by Y. The model we have to estimate is thus

$$C_t = a + bC_{t-1} + cY_t + e_t$$

where t denotes the time period. Our ordinary least-squares estimators are

$$\hat{a} = \bar{C} - \hat{b}\bar{C}_{-1} - \hat{c}\bar{Y}$$

$$\hat{b} = \frac{\text{Cov}(C_{-1}, C)\,\text{Var}\,Y - \text{Cov}(Y, C)\text{Cov}(C_{-1}, Y)}{\text{Var}\,C_{-1}\,\text{Var}\,Y - [\text{Cov}(C_{-1}, Y)]^2}$$

$$\hat{c} = \frac{\text{Cov}(Y, C)\text{Var}\,C_{-1} - \text{Cov}(C_{-1}, C)\text{Cov}(C_{-1}, Y)}{\text{Var}\,C_{-1}\,\text{Var}\,Y - [\text{Cov}(C_{-1}, Y)]^2}$$

(Cov and Var both refer to sample values and the subscript -1 indicates a lag of one time period.) The necessary calculations of the components of these estimators are given in the table overleaf.

$$\text{Cov}(C_{-1}, C) = \Sigma C_{-1}C/N - \bar{C}_{-1}\bar{C} = 466.137 - (21.8)(21.29) \quad = 2.015$$

$$\text{Var}\,Y \qquad = \Sigma Y^2/N - \bar{Y}^2 \qquad = 567.038 - 23.76^2 \qquad = 2.5004$$

$$\text{Cov}(Y, C) \quad = \Sigma YC/N - \bar{Y}\bar{C} \qquad = 520.190 - (23.76)(21.8) \quad = 2.222$$

$$\text{Cov}(C_{-1}, Y) = \Sigma C_{-1}/N - \bar{C}_{-1}\bar{Y} \quad = 508.103 - (21.29)(23.76) = 2.2526$$

$$\text{Var}\,C_{-1} \qquad = \Sigma C_{-1}^2/N - \bar{C}_{-1}^2 \qquad = 455.345 - 21.29^2 \qquad = 2.0809$$

Thus

$$\hat{b} = \frac{(2.015)(2.5004) - (2.222)(2.2526)}{(2.0809)(2.5004) - 2.2526^2} = 0.256$$

$$\hat{c} = \frac{(2.222)(2.0809) - (2.015)(2.2526)}{(2.0809)(2.5004) - 2.2526^2} = 0.658$$

The two coefficients that we have obtained seem fairly plausible as they both lie between zero and unity. However, it is noticeable that it is very important to use accurate arithmetic as the denominator in question for both coefficients is very close to zero. Thus any slight change in the data might alter the value of the coefficients fairly substantially. However, we would tend to expect the relationship to be fairly stable. $\hat{c}$ is still our marginal propensity to consume and $\hat{b}$ the effect of habits. Obviously we have some difficulty in distinguishing the two effects as C_{t-1} and Y_t are closely correlated.

	C	C_{-1}	Y	C^2	C^2_{-1}
1962	19.3	18.9	20.9	372.49	357.21
1963	20.1	19.3	21.8	404.01	372.49
1964	20.8	20.1	22.7	432.64	404.01
1965	21.2	20.8	23.2	449.44	432.64
1966	21.6	21.2	23.7	466.56	449.44
1967	22.1	21.6	24.1	488.41	466.56
1968	22.7	22.1	24.6	515.29	488.41
1969	22.8	22.7	24.7	519.84	515.29
1970	23.4	22.8	25.6	547.56	519.84
1971	24.0	23.4	26.3	576.00	547.56
TOTAL	218.0	212.9	237.6	4,772.24	4,553.45
TOTAL/N	21.8	21.29	23.76	477.224	455.345

	Y^2	CC_{-1}	CY	$C_{-1}Y$
1962	436.81	364.77	403.77	395.01
1963	475.24	387.93	438.18	420.74
1964	515.29	418.08	472.16	456.27
1965	538.24	440.96	491.84	482.56
1966	561.69	457.92	511.92	502.44
1967	580.81	477.36	532.61	520.56
1968	605.16	501.67	558.42	543.66
1969	610.09	517.56	563.16	560.69
1970	655.36	533.52	599.04	583.68
1971	691.69	561.60	631.20	615.42
TOTAL	5,670.38	4,661.37	5,201.90	5,081.03
TOTAL/N	567.038	466.137	520.190	508.103

2.   We can start by estimating b and c by ordinary least squares.

$$\hat{b} = \frac{\Sigma yv\Sigma r^2 - \Sigma rv\Sigma ry}{\Sigma y^2 \, \Sigma r^2 - (\Sigma yr)^2}$$

and

$$\hat{c} = \frac{\Sigma rv\Sigma y^2 - \Sigma yv\Sigma ry}{\Sigma y^2 \, \Sigma r^2 - (\Sigma yr)^2}$$

where small letters denote deviations of variables from their sample means. Thus

$$\hat{b} = \frac{(6.24)(76.36) - (11.0)(35.07)}{(19.2)(76.36) - (35.07)^2} = 0.384$$

$$\hat{c} = \frac{(11.0)(19.2) - (6.24)(35.07)}{(19.2)(76.36) - (35.07)^2} = -0.0323$$

$$\hat{\sigma}_{\hat{b}} = \frac{\hat{\sigma}}{\sqrt{(\Sigma r^2 - \Sigma ry/\Sigma y^2)}} = \frac{\sqrt{0.0032}}{\sqrt{(76.36 - 35.07/19.2)}} = 0.00655$$

$$\hat{\sigma}_{\hat{c}} = \frac{\hat{\sigma}}{\sqrt{(\Sigma y^2 - \Sigma ry/\Sigma r^2)}} = \frac{\sqrt{0.0032}}{\sqrt{(19.2 - 35.07/76.36)}} = 0.0131$$

At the 5% level of significance $t(38) = 2.02$

$t_{\hat{b}} = 0.384/0.00655 = 58.63$

$t_{\hat{c}} = -0.0323/0.0131 = -2.47$

we thus reject the null hypothesis in both cases. If we had used the 1% level we would have accepted the null hypothesis for c.

Both coefficients have plausible signs and magnitudes. As GDP rises so does investment, but by a smaller amount, and as interest rates rise investment falls and vice-versa. It is also worth noting that the relationship with interest rates is relatively weak — a fact which also coincides with our expectations.

3. (i) Clearly our first step is to test the null hypothesis that each of the individual coefficients is equal to zero. Let us choose 5% as our significance level in these tests. We can obtain the appropriate t statistics under the null hypothesis that the individual coefficients are zero by dividing the estimated coefficient by its standard error.

Policy 'off'		Policy 'on'	
$t(27)_{0.05} = 2.05$		$t(33)_{0.05} = 2.03$	
$t_{\hat{a}} = 5.79$	reject H_0	$t_{\hat{a}} = 2.27$	reject H_0
$t_{\hat{b}} = -3.64$	reject H_0	$t_{\hat{b}} = -0.56$	accept H_0
$t_{\hat{c}} = 6.25$	reject H_0	$t_{\hat{c}} = 0.93$	accept H_0
$t_{\hat{d}} = 0.07$	accept H_0	$t_{\hat{d}} = 1.61$	accept H_0

(ii) We observe coefficients which conform to our expectations. If unemployment is high then labour is easier to obtain and hence wages will not rise so fast. If the rate of increase in prices is high then wages will tend to rise faster to offset this. (Note that the causation works in the opposite direction as well.) Lastly, if employees are better organized their bargaining power may improve however, this hypothesis is not borne out by the data.

(iii)

$$F(3,27)_{0.05} = 2.96, \qquad F(3,33)_{0.05} = 2.89$$

$$F = (R^2/k)/[(1 - R^2)/(n - k - 1)]$$

where k is the number of exogenous variables (excluding the constant and n is the number of observations.

$F_{\text{policy off}} = 53.5$ reject H_0, $\qquad F_{\text{policy on}} = 1.76$ accept H_0

(iv) If we assume that if an incomes policy had not been imposed during the policy 'on' periods the relationship would have been the same as in the policy 'off' periods then clearly an incomes policy has had an effect. The relationships between unemployment and the rate of change of prices, and the rate of change of wages are no longer significant. Both coefficients have also moved nearer zero. Thus if the changes are the result of the incomes policies then they have to an extent 'worked'. The rate of increase in wages may not be lower however as the two changes in coefficients may offset each other. In fact the fall in the constant term shows a lower general rate of increase in wages when the policy is 'on'.

4.   The general tendency would be to expect earnings to rise with the amount of education, and for male employees to earn more than female employees. If we wish to formulate the model so as to estimate interaction effects as well as the effects of the individual states, we must have six variables as there are six possible states: F, $F U$, $F P$, M, $M U$, $M P$.

Let us define the model in the form

$$Y = b_1 X_1 + b_2 X_2 + b_3 X_3 + b_4 X_4 + b_5 X_5 + b_6 X_6$$

There are clearly several possible choices of the Xs such that we can estimate the required results. The following is one such schema

X_1 = 1 for all employees
X_2 = 1 if the person is male and 0 if female
X_3 = 1 if the person has a university first degree only, 0 otherwise
X_4 = 1 if the person has a postgraduate degree, 0 otherwise
X_5 = 1 if the person is male *and* has a university first degree only 0 otherwise
X_6 = 1 if the person is male *and* has a postgraduate degree, 0 otherwise

The first step is thus to construct the variables to be used in the regression, and this is shown for the first few questionnaires.

Q	Y	X_1	X_2	X_3	X_4	X_5	X_6
1	2,500	1	1	0	1	0	1
2	1,400	1	0	0	0	0	0
3	1,600	1	0	1	0	0	0
4	2,800	1	1	0	1	0	1
5	2,600	1	1	1	0	1	0
—	.	.	.	.	.	.	.
—	.	.	.	.	.	.	.
32	2,000	1	1	0	0	0	0

Since we have six exogenous variables it will be very tedious to estimate the regression either by hand or by calculating machine. Before proceeding to the computer solution, we can check the values of the matrix of sums of squares and cross-products. Thus the first row of the matrix is

$$\sum_{i=1}^{n} Y_i^2, \quad \sum_{i=1}^{n} X_{i1} Y_i, \quad \sum_{i=1}^{n} X_{i2} Y_i, \ldots$$

the second is

$$\sum_{i=1}^{n} Y_i X_{i1}, \quad \sum_{i=1}^{n} X_{i1}^2, \quad \sum_{i=1}^{n} X_{i2} X_{i1}, \ldots$$

and so on, where n is the total number of questionnaires and i the individual questionnaire and as an example

$$\Sigma Y_i^2 = 2{,}500^2 + 1{,}400^2 + 1{,}600^2 + \ldots + 2{,}000^2$$
$$= (16{,}065)(10^4)$$

The matrix is thus symmetric, the diagonal elements are sums of squares and the second row (and second column) gives the sums of each variable as $X_1 = 1$ for all observations.

	Y	X_1	X_2	X_3	X_4	X_5	X_6
Y	16,065	691	471	312	165	194	137
X_1	691	32	20	13	6	7	5
X_2	471	20	20	7	5	7	5
X_3	312	13	7	13	0	7	0
X_4	165	6	5	0	6	0	5
X_5	194	7	7	7	0	7	0
X_6	137	5	5	0	5	0	5

Note that Y is divided by 100 for ease of calculation. The results of the regression (Y has original values) are

$$Y = 1{,}500X_1 + 271.4X_2 + 466.7X_3 + 1{,}300X_4 + 533.3X_5 - 331.4X_6$$
$$\phantom{Y = 1{,}500X_1 + } (177.0) \quad (183.7) \quad (343.6) \quad (250.3) \quad (390.9)$$

$R^2 = 0.752, \quad F = 15.8$

The coefficients of X_3, X_4 and X_5 are significantly different from zero at 5% as $t(25)_{0.05} = 2.06$. ($F(5,26)_{0.05} = 2.59$.)
Total expected earnings for females without any degree (b_1) is £1,500.
Total expected earnings for males without any degree ($b_1 + b_2$) is £1,771.4.
Total expected earnings for females with a first degree ($b_1 + b_3$) is £1,966.7.

Total expected earnings for males with a first degree ($b_1 + b_2 + b_3 + b_5$) is £2,771.4.

Total expected earnings for females with a postgraduate degree ($b_1 + b_4$) is £2,800.

Total expected earnings for males with a postgraduate degree ($b_1 + b_2 + b_4 + b_6$) is £2,740.

The effect of having a first degree for males compared to no degree is

$$b_3 + b_5 = £1,000$$

The effect of having a first degree for females compared to no degree is

$$b_3 = £466.7$$

Thus the effect of being male among first-degree holders is

$$b_5 = £533.3$$

The effect of having a postgraduate degree for males compared to no degree is

$$b_4 + b_6 = £968.6$$

The effect of having a postgraduate degree for females compared to no degree is

$$b_4 = £1,300$$

Thus the effect of being male among postgraduates degree holders is

$$b_6 = -£331.4$$

The effect of having a postgraduate degree for males compared to a first degree is

$$b_4 + b_6 - b_3 - b_5 = -£31.4$$

The effect of having a postgraduate degree for females compared to a first degree is

$$b_4 - b_3 = £833.3$$

The effect of being male for those without degrees is

$$b_2 = £271.4$$

The two points of interest which do not conform to our original hypothesis are (1) The slight negative effect of a further degree for males — possibly because they have only been earning for a short period — (2) the positive effect of being female among postgraduates — which may be abnormal as we only have one observation.

5. (i) It is clear from a quick glance at both the original data and Figure 7.2 that the series has a yearly cycle round a rising trend during the period of the observations.

Table 7.2

		(1) M3	(2) LS trend	(3) (1) − (2)	(4) seas. adj. series	(5) Σ4	(6) Σ8	(7) trend	(8) (1) − (7)	(9) seas. adj. series
1968	IV	649	−157.6	806.6	100					77
1969	I	−301	−61.2	−239.8	281	495				300
	II	−52	35.2	−87.2	−121	503	998	124.75	−176.75	−217
	III	199	131.6	67.4	235	374	877	109.63	89.37	334
	IV	657	228.0	429.0	108	1,159	1,533	191.63	465.37	85
1970	I	−430	324.4	−754.4	152	1,347	2,506	313.25	−743.25	171
	II	733	420.8	312.2	664	1,586	2,933	366.63	366.67	568
	III	387	517.2	−130.2	423	2,031	3,617	452.13	−65.13	522
	IV	896	613.6	282.4	347	1,769	3,800	475.00	421.00	324
1971	I	15	710.0	−695.0	597	1,834	3,603	450.38	−435.38	616
	II	471	806.4	−335.4	402	2,366	4,200	525.00	−54.00	306
	III	452	902.8	−450.8	488	2,811	5,177	647.13	−195.13	587
	IV	1,428	999.2	428.8	879	4,038	6,849	856.13	571.87	856
1972	I	460	1,095.6	−635.6	1,042	4,541	8,579	1,072.38	−612.38	1,061
	II	1,698	1,192.0	506.0	1,629	5,299	9,840	1,230.00	468.00	1,533
	III	955	1,288.4	−333.4	991	5,735	11,034	1,379.25	−424.25	1,090
	IV	2,186	1,384.8	801.2	1,637	5,565	11,300	1,412.50	773.50	1,614
1973	I	896	1,481.2	−585.2	1,478	6,955	12,520	1,565.00	−669.00	1,497
	II	1,528	1,577.6	−49.6	1,459					1,363
	III	2,345	1,674	671.0	2,381					2,480

L.S. adjustment

	Q1	Q2	Q3	Q4	Σ
1968				806.6	
1969	−239.8	−87.2	67.4	429.0	
1970	−754.4	312.2	−130.2	282.4	
1971	−695.0	−335.4	−450.8	428.8	
1972	−635.6	506.0	−333.4	801.2	
1973	−585.2	−49.6	671.0		
Σ	−2,910.0	346.0	−176.0	2,748.0	
Σ/5	−582.0	69.2	−35.2	549.6	1.6
Adj.	−0.4	−0.4	−0.4	−0.4	0
	−582.4	68.8	−35.6	549.2	
	−582	69	−36	549	

M.A. adjustment

	Q1	Q2	Q3	Q4	Σ
1968					
1969		−176.75	89.37	465.37	
1970	−743.25	366.67	−65.13	421.00	
1971	−435.38	−54.00	−195.13	571.87	
1972	−612.38	468.00	−424.25	773.50	
1973	−669.00				
Σ	−2,460.01	603.92	−595.14	2,231.74	−54.87
Σ/4	−615.00	150.98	−148.79	557.94	13.72
Adj.	13.72	13.72	13.72	13.72	13.72
	−601.28	164.70	−135.07	571.66	0.01
	−601	165	−135	572	

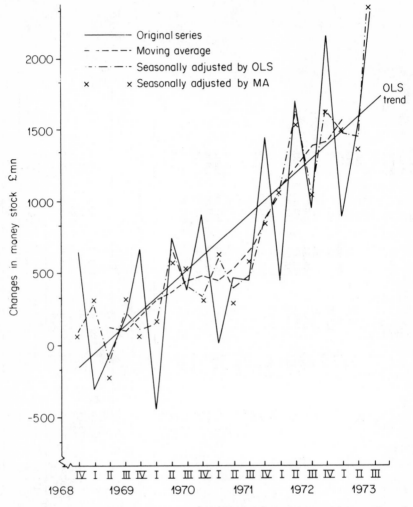

Figure 7.2   Changes in money stock $M3$, quarterly, U.K. Source: *Financial Statistics*

(ii) We can see from Figure 7.2 that a linear trend approximates the data fairly closely.

(iii) If we fit a linear trend by least squares we obtain the equation

$$M3 = -254 + 96.4t, \quad R^2 = 0.56$$
$$(19.1)$$

if we also fit a set of seasonal dummies

$$M3 = 368 + 91.6t - 1{,}156Q_1 - 500Q_2 - 786Q_3$$
$$(9.42) \quad (147) \quad (147) \quad (157)$$

$R^2 = 0.91$ and all coefficients are significant at 5%

However, we could calculate the seasonal dummies by subtracting the trend from $M3$ and averaging the quarterly deviations from the trend and this is shown in columns (1)—(4) in the table overleaf. The average deviations are, $Q1$ −582, $Q2$ 69, $Q3$ −36 and $Q4$ 549. As can be seen from Figure 7.2, when we subtract these constants from the original series to form the seasonally adjusted series, column (4), the result still fluctuates widely round the trend.

The second half of the table shows the calculation of the four-quarter moving average and the average seasonal deviations from that, $Q1$ −601, $Q2$ 165, $Q3$ −135 and $Q4$ 572. The first point to notice is that the moving average trend lies below the linear trend in 1970 and 1971, perhaps indicating that a logarithmic trend might be a good approximation. Secondly the seasonal constants are all larger in absolute value than those derived from the linear trend. This second seasonally adjusted series, column (9) in the table, also fluctuates more widely than the first series. However, despite the differences in trend the two adjusted series are sufficiently close together that it is not possible to draw them distinctly on the graph.

(iv) The seasonal pattern of changes in the money stock is reasonably clear. The major expansion occurs in the fourth quarter of the year, and is perhaps related to increased consumer spending and consequent increase in credit. Similarly the decrease in the first quarter of the year may be related to the end of the financial year and the payment of tax bills.

8 The Role of Econometrics

8.1 Problems Imposed by Reality

During the last three chapters we have steadily developed an estimation method for a specific form of linear model. The desirable properties of the particular least-squares estimators stem from the assumptions of the model and if these assumptions do not hold then we must adjust our estimation procedure. It is an unfortunate fact that economic life often does not conform to the pattern of our simple model; the subject of econometrics is centred on the problems of specification, estimation, testing and forecasting with models which will represent economic behaviour more closely. In this chapter we shall look at two general areas. Firstly we shall consider the difficulties which stem from the fact that we cannot always isolate economic relationships; events happen simultaneously and we cannot assume that other variables are held constant for the purposes of our analysis. We must, therefore, 'identify' the individual relationships and consider methods to estimate them simultaneously. Secondly, we shall turn our attention to occasions where, although we have only a single equation, some of the assumptions of our basic model are no longer realistic.

This chapter thus forms the heart of our book. In it we are showing how our statistical methodology can cope with the many convolutions and intricacies of economic reality.

8.2 Simultaneous Events

While it is very easy for economists to talk about simple relationships for the purpose of analysis, it is by no means equally easy to estimate such relationships in every case, even if the economists' hypotheses are completely correct. Let us take a simple example from supply and demand. In a particular market the supply function is

$$Q = a + bP \tag{1}$$

where Q stands for quantity and P for the price of the product; a is an arbitrary constant and $b > 0$. The demand function in the market is

$$Q = c + dP \tag{2}$$

where $c > a$ and $c > 0$ and $d < 0$. Let us also assume that when the quantity demanded is equal to the quantity supplied the price at which the commodity is traded is positive. Thus we have the situation shown in Figure 8.1.

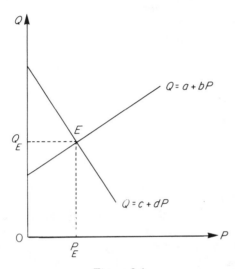

Figure 8.1

If the market is perfect it will always trade at equilibrium, E, where supply and demand are equal. Thus whenever we looked at the transactions in the market we would observe the same price, P_E, and the same quantity Q_E. We could not therefore estimate either (1) or (2) by any method whatsoever from our observations on the market as they all have the value E and there is an infinite number of pairs of lines intersecting at E, all of which would produce the same observed values Q_E and P_E.

Thus at least one relationship must vary so that we obtain more than one value of Q and P if we are to be able to estimate anything at all. Let us say that it is the demand relationship that varies

$$Q_i = c + dP_i + e_i \tag{3}$$

where e has different values for each observation, i. You will notice that at this stage e does not have to be our random error term; it can have any properties consistent with positive quantities and prices. However, let us say for simplicity that $E[e] = 0$. Now, when we observe the quantities and prices at which the commodity was traded, we will

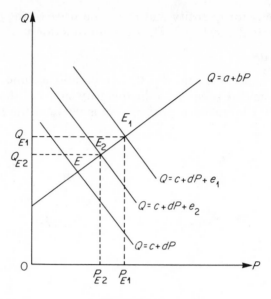

Figure 8.2

obtain a set of points on the supply curve because (1) still holds. If we consider two periods 1 and 2, where, let us say, $e_1 > e_2 > 0$, the situation we observe is shown in Figure 8.2.

The market is in equilibrium in each period as we assumed that it was perfect, so we observe Q_{E1}, P_{E1} in the first period and Q_{E2}, P_{E2} in the second period. This perhaps seems odd, but if demand fluctuates and the supply curve remains constant we can estimate the supply curve and not the demand curve. In the same way, if it is the supply curve which varies and the demand curve which is constant it is the demand curve we can estimate.

In reality, however, both supply *and* demand curves are likely to fluctuate, so that we do not observe a set of points on either the demand or supply functions. Firstly let us formulate our supply function to include the fluctuating element, say u_i, where $E[u] = 0$

$$Q_i = a + bP_i + u_i \tag{4}$$

For every period i we will have both an e_i and a u_i; thus the resulting observed values of P and Q will be those shown in Figure 8.3, for three such is. The points of intersection of the functions, E_1, E_2 and E_3, are emphasized to distinguish them clearly from the other points of intersection which are not observable since the intersecting lines do not refer to the same time period. With only three points, E_1, E_2 and E_3, we do not have enough degrees of freedom to make it sensible to estimate a line to which they correspond. However, in Figure 8.4 we have shown a more general situation where we have a much larger

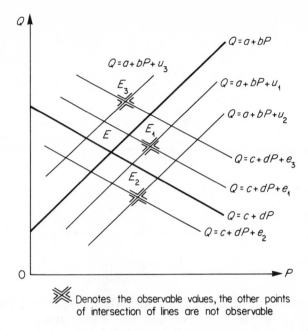

Figure 8.3

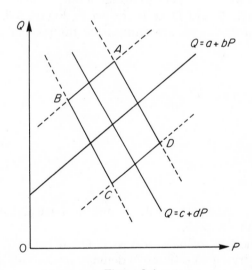

Figure 8.4

number of observations. All observations occur within the quadrilateral $ABCD$ where the sides of the quadrilateral represent the most extreme fluctuations of the two functions. Thus BA is a section of the supply function with the largest u_i and CD a section of the supply function with the smallest u_i (in this latter case u_i is clearly negative as CD lies

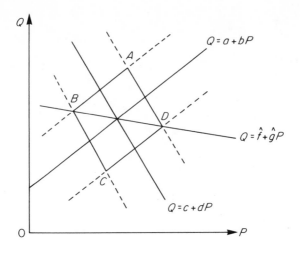

Figure 8.5

below the supply function). Since we have not specified the distribution of the e_i or the u_i, all we can say is that the observations are scattered throughout the quadrilateral $ABCD$. If we now regress Q on P by least squares we can expect to obtain a line similar to that which we can draw through B and D as is shown in Figure 8.5. In calculating least-squares regressions we are minimizing the residual sum of squares

$$\sum_i \hat{v}_i^2$$

subject to

$$\sum_i \hat{v}_i = 0$$

in the equation

$$Q_i = \hat{f} + \hat{g}P_i + \hat{v}_i \tag{5}$$

with respect to $\hat{f}$ and $\hat{g}$, where the $\hat{v}_i$ are the vertical distances between the observed points and (5).

However, although we have regressed Q on P in (5) it is not clear whether we are trying to estimate a demand or a supply function. Both (1) and (2) have the same linear form as (5) and of course as each other, but (5) does not look like either of them, in fact it looks very much as if $\hat{g}$ might be zero. In other words our estimated line would show that there is no relation between quantity and price. However, this latter result stems from the particular $ABCD$ that we have drawn; we could draw many others, but the estimated line, (5), will have the properties that $c > \hat{f} > a$ and $b > \hat{g} > d$. Using the data from the first two

Table 8.1
Values of Variables in the simple market model

P	Q	Y	W	P^2	$Q^2 \times 10^2$	$Y^2 \times 10^4$	W^2
10	200	1,200	31.6	100	400	144	998.56
12	180	1,100	28.0	144	324	121	784
14	190	1,150	26.8	196	361	132.25	718.24
11	200	1,250	31.5	121	400	156.25	992.25
12	205	1,300	31.5	144	420.25	169	992.25
15	210	1,400	30.0	225	441	196	900
16	200	1,300	27.3	256	400	169	745.29
18	180	1,200	23.7	324	324	144	561.69
18	175	1,100	22.0	324	306.25	121	484
20	180	1,150	20.8	400	324	132.25	432.64
$\Sigma 146$	1,920	12,150	273.2	2,234	3,700.5	1,484.75	7,608.92

PQ	$PY \times 10^2$	PW	$QY \times 10^4$	$QW \times 10^2$	$YW \times 10^2$	$\hat{P}$
2,000	120	316	24.0	63.2	379.2	10.074
2,160	132	336	19.8	50.4	308.0	11.987
2,660	161	375.2	21.85	50.92	308.2	14.026
2,200	137.5	346.5	25.0	63.0	393.75	11.015
2,460	156	378	26.65	64.58	409.5	11.856
3,150	210	450	29.4	63.0	420.0	15.036
3,200	208	436.8	26.0	54.6	354.9	16.050
3,240	216	426.6	21.6	42.66	284.4	17.962
3,150	198	396	19.25	38.5	242.0	17.977
3,600	230	416	20.7	37.44	239.2	20.017
$\Sigma 27,820$	1,768.5	3,877.1	234.25	528.30	3,339.15	

columns of Table 8.1, for example, we obtain

$$Q = 162 - 2.07P$$
$$(16.3)\ (1.09)$$

and thus accept the null hypothesis that there is no relationship between price and quantity, when in fact both supply and demand are related to price.

Whatever line we do estimate in (5) we have neither a demand function nor a supply function. With the information that we have we cannot 'identify' either of the two functions. When only one function was allowed to fluctuate we could only estimate the other function and not the fluctuating one. Thus, in the terminology of 'identification', the constant function is 'identified' and the fluctuating one is 'not identified'.

The reason that we cannot identify the equations is that we have four unknowns, a, b, c and d, yet when we estimate (5) by least squares we have only two Normal Equations. Thus the original system is insoluble.

Furthermore we can show that estimating either (3) or (4) by (5) violates the assumption that P and the error terms are uncorrelated. If we take the case of (3) as an example, by equating (3) and (4) we have

$$c + dP + e = a + bP + u$$

and hence

$$P = \frac{a - c}{d - b} + \frac{u - e}{d - b}$$

Taking expectations

$$E[P] = \frac{a - c}{d - b}$$

as $E[u] = E[e] = 0$. Thus we can find the covariance of e and P

$$\text{Cov}(e, P) = E[e(P - E[P])]$$

$$= E\left[e\frac{u - e}{d - b}\right]$$

$$= \frac{1}{d - b}\{E[eu] - E[e^2]\}$$

$$= \frac{1}{b - d}E[e^2] \quad \text{as} \quad E[eu] = 0$$

We can see that $E[e^2]/(b - d)$ is not equal to zero and thus p and e are correlated.

8.3 The Identification Problem

We have shown in the very simple case of supply and demand the difficulties of simultaneous equations. If we are to tackle this as a general problem and show its solution we must formulate the system in a general way rather than just considering one specific instance. The problem in the case of demand and supply was that both price and quantity were determined by the system. In other words they are both endogenous variables. In general we can say that all models are composed of equations in endogenous and exogenous variables. Let us refer to endogenous variables by X and exogenous variables by Z. We could therefore write down every equation in the model in the form

$$a_{i1}X_1 + a_{i2}X_2 + \ldots + b_{i1}Z_1 + b_{i2}Z_2 + \ldots + e_i = 0 \tag{6}$$

where i is the number of the equation. In fact every equation can have exactly the same form if we allow some of the coefficients a and b to be zero. For example, if the demand and supply equations in our previous model are modified to include, say, personal disposable income, Y, and the weather, W, respectively

$$Q = c + dP + hY + e \tag{7}$$

$$Q = a + bP + jW + u \tag{8}$$

We can write them both as

$$a_{11}Q + a_{12}P + b_{11}Y + b_{12}W = e_1 \qquad (9)$$

$$a_{21}Q + a_{22}P + b_{21}Y + b_{22}W = e_2 \qquad (10)$$

where $b_{12} = b_{21} = 0$, $a_{11} = a_{21} = 1$, $a_{12} = -d$, $a_{22} = -b$ and so on.

Equations (7) and (8) are the structure of our model. They have the form that our theory suggests. But if we wish to estimate the model we want to determine the values of each of our endogenous variables. We cannot merely go ahead and estimate each of the structural equations by least squares, because this will give us two conflicting estimates of Q and no estimate at all of P. Thus we want two equations, the first with Q on the left-hand side, and the second with P on the left-hand side. Firstly, we can substitute in (8) for Q using (7).

$$c + dP + hY + e = a + bP + jW + u$$

Collecting terms

$$(d - b)P = (a - c) + jW - hY + (u - e)$$

we obtain

$$P = \frac{a - c}{d - b} + \frac{j}{d - b} W - \frac{h}{d - b} Y + \frac{u - e}{d - b} \qquad (11)$$

Similarly substituting in (7) for P using (8)

$$Q = c + d\left(\frac{Q - a - jW - u}{b}\right) + hY + e \qquad (12)$$

we obtain

$$Q = \left(\frac{cb - ad}{b - d}\right) - \frac{dj}{b - d} W + \frac{bh}{b - d} Y + \frac{eb - du}{b - d} \qquad (13)$$

These two equations, (11) and (13), are known as the *reduced form* of the model, and we can rewrite them as

$$P = \alpha_1 + \beta_1 W + \gamma_1 Y + v_1 \qquad (11a)$$

$$Q = \alpha_2 + \beta_2 W + \gamma_2 Y + v_2 \qquad (13a)$$

where

$$\alpha_1 = \frac{a - c}{d - b} \qquad (14)$$

$$\beta_1 = \frac{j}{d - b} \qquad (15)$$

and so on. If we estimate (11a) and (13a) by ordinary least squares we will obtain values for α_1, β_1, γ_1, α_2, β_2 and γ_2 and thus derive six equations, of which (14) and (15) are the first two, in the six unknowns a, b, c, d, h and j. However, we can only solve the estimated reduced form to find the coefficients of the structural equations if the model is identified.

★ As in the case of multiple regression we can express the relation
★ between the structure and the reduced form much more simply in
★ matrix algebra. If we write the structure as shown in (6) as
★
★ $$AX + BZ = 0 \tag{6a}$$
★
★ where all equations in the model are shown in (6a), then the
★ reduced form is simply
★
★ $$X = -A^{-1}BZ \tag{6b}$$

The data for our example are set out in Table 8.1, and when we estimate (11a) and (13a) by OLS we obtain

$$P = 21.441 + 0.01682Y - 0.9885W, \quad R^2 = 0.9997$$
$$\quad\quad (0.288) \quad (0.00029) \quad (0.0069)$$
$$Q = 55.050 + 0.07870Y + 1.5129W, \quad R^2 = 0.944$$
$$\quad\quad (14.035) + (0.01392) \quad (0.3375)$$

Solving for the structure we obtain

$$Q = \quad 87.5 - 1.52P + 0.104Y$$
$$Q = -45.3 + 4.68P + 6.18W$$

★ If the reader wishes to check this he should firstly obtain the six
★ equations between the reduced form coefficients and the structural
★ coefficients and then solve them; however, this is not a very simple
★ operation.

The rules concerning the determination of the degree of identification of the model are quite simple. Firstly a model is not logically complete if it has less structural equations than there are endogenous variables. If there are too few equations then we will have more unknowns in the structural equations than we have equations. For example, if we had tried to estimate (3) without acknowledging the existence of (4), we would be unable to estimate the structural parameters as the estimated equation would in fact still be (5).

The second necessary condition is that there must be sufficient restrictions on the coefficients of the equations to distinguish between them. Thus although we can write (7) and (8) in the same form as each other as in (9) and (10), (9) has the restriction that $b_{12} = 0$ and (10) has the restriction that $b_{21} = 0$. Equations (3) and (4), on the other hand, are indistinguishable. Thus, in that case, although we have as many equations as endogenous variables in the system and thus fulfil the first condition the model does not fulfil the second condition. In effect if there are K equations in the system we must have at least $K - 1$ restrictions.

The third condition is that each structural equation must have at least $K - 1$ variables with coefficients restricted to zero, where K is the number of equations. Clearly (9) and (10) exactly meet this condition as $K - 1 = 1$ in that case. This system is thus exactly or 'just'-identified. If equation (7) is replaced by equation (3) where b_{11} is also constrained equal zero the system is 'over'-identified as only one constraint is necessary, but by our third condition an overidentified system is still 'identified'. It is only the remaining case where the number of constraints is insufficient, for example if both personal disposable income and the weather effect demand, that the system is 'under'-identified and therefore not estimable.

8.4 An Introduction to Simultaneous Equation Methods

We have distinguished three different situations: overidentification, exact identification and underidentification. We have established that an underidentified model cannot be fully estimated, although it may be possible to identify and estimate some of the equations in the model. We have also stated that an exactly identified system can be estimated by the use of our ordinary least-squares method on the reduced form of the model. This method is usually known as *Indirect Least Squares*. The overidentified case on the other hand presents a rather difficult situation where the most useful method of estimation varies with the amount of information that we have on the restrictions on the model. In this section we shall explain an estimation method for simple cases and indicate the approach of some of the other methods.

For simplicity we shall proceed with our simple demand and supply model shown in equations (7) and (8). It will be remembered that this model is in fact exactly identified and not overidentified, but the method is still valid in this case and saves us the complication of specifying a third exogenous variable in order to achieve overidenti-fication. As we saw (pp. 175—6) the problem in estimating (7) and (8) is that P is correlated with the error term in each equation. We can overcome this problem by firstly removing the correlation and then estimating the relationship. We thus have a two-stage procedure. The source of the correlation between P and the error term is the fact that P is related to both Y and W as we saw in equation (11a). If we estimate (11a) by least squares and calculate the estimated values of P for each pair of Y and W, we are in fact removing the correlated element from P. These estimated values $\hat{P}$ are obtained from

$$\hat{P} = \alpha_1 + \beta_1 W + \gamma_1 Y \tag{16}$$

180

so that from (11) and (11a)

$$P - \hat{P} = v_1 = \frac{u - e}{d - b}$$ (17)

If we now substitute $\hat{P}$ for P in (7) we would be rewriting our model as

$$Q = c + d\hat{P} + hY + \left[\frac{d(u - e)}{d - b} + e \right]$$ (18)

and similarly for equation (8)

$$Q = a + b\hat{P} + jW + \left[\frac{b(u - e)}{d - b} + u \right]$$ (19)

$\hat{P}$ is now uncorrelated with the two new composite error terms (shown in square brackets), so we can use our least-squares method to estimate (18) and (19) without violating the assumptions behind its correct use.

We estimated (11a) on p. 178, and the values of $\hat{P}$ derived from (16) are shown in the last column of Table 8.1. If we now estimate (18) and (19) we obtain

$$Q = 87.5 - 1.52\hat{P} + 0.104Y$$

and

$$Q = -45.3 + 4.68\hat{P} + 6.18W$$

The first point to notice is that the estimated coefficients are identical to those obtained by Indirect Least Squares on p. 178. This perhaps can be seen as a form of validation for our method, although, of course, when the model is overidentified we can no longer use Indirect Least Squares so the comparison is then not possible. As we have seen, our two-stage procedure uses Least Squares twice; it is therefore known as *Two-Stage Least Squares* (TSLS or 2SLS for short).

However, 2SLS is only one among several simultaneous equation methods. Although these methods may be equivalent as the sample size tends to infinity they have different ill-defined small sample properties. In practice, choice is therefore exercised over which method to use. We have explained 2SLS not because it is the 'best' method, but because it is one which is very widely used, is straightforward, does not require more information than we have shown and is consistent.

Further methods can be divided into two very general classes: those like 2SLS, which can be applied to each individual equation in the model in turn, and those which are applied to the whole system at one step. All these further methods incorporate information about the sampling variances and covariances of the structural coefficients in their estimation procedure (as does 2SLS). The difference between the two categories of methods is merely that in the first case, where equations can be treated individually, the information is separable.

8.5 Single-Equation Problems: Multicollinearity

Having given brief consideration to the difficulties that simultaneity imposes on our model, we shall now examine some of the problems that reality imposes on our models where single equations are appropriate. The first of these problems concerns the interrelationship of the exogenous variables in the model. This problem is known as multicollinearity, and it has two facets. In the first case if there is an exact linear relationship between the exogenous variables in the model the normal equations are insoluble. In the second and more important case when there is a fairly close but not exact linear relationship between the exogenous variables the least-squares estimates may not appear unusual, but in fact they are liable to inaccuracy of calculation, their standard errors will tend to be very large and the values are very sensitive to changes in the data. There is a very real danger that in this latter case an extremely erroneous inference will be drawn from the calculated statistics. We must be able to diagnose the existence of this state of affairs and have appropriate courses of action to take should it exist.

However, let us begin by examining the first case of an exact linear relation. On p. 140 we set out the form of the least-squares estimators in the simplest case of multiple regression. In the equation $Y = b_0 + b_1 X_1 + b_2 X_2$ both b_1 and b_2 have the same denominator, $\text{Var } X_1 \text{ Var } X_2 - [\text{Cov}(X_1 X_2)]^2$. If X_1 and X_2 are perfectly correlated

$$r_{X_1 X_2} = 1 = \frac{\text{Cov}(X_1 X_2)}{\sqrt{(\text{Var } X_1 \text{ Var } X_2)}}$$

or rearranging

$$\text{Cov}(X_1 X_2) = \sqrt{(\text{Var } X_1 \text{ Var } X_2)}$$

Thus the denominator of both b_1 and b_2 is zero and both estimators do not have finite values. This same result is obtained both when there are more variables in the regression equation and when the linear relation is between more than two variables.

★ In matrix terms, if X is the matrix of regressor variables, then if
★ there is a linear relationship between the variables, $|X'X| = 0$ and
★ $[X'X]^{-1}$ does not exist, so we cannot calculate the least-squares
★ coefficients $\hat{b} = [X'X]^{-1} X'y$.

In the case of a close but not exact linear relation between the regressor variables the source of inaccuracy is easy to see as the denominator in b_1 and b_2 will be close to zero and any small alterations in the values of its components may have substantial effects on the coefficients. Secondly, from p. 142 we can see that the same factor affects the standard errors of b_1 and b_2; the greater the correlation between X_1 and X_2 the larger the standard error. However, the really interesting problems are those of recognition of multicollinearity and the appropriate action to be taken if multicollinearity exists. Recognition in the first place can stem from the effect on standard errors. If R^2 indicates the existence of a highly significant relation between the Xs and Y but none of the t statistics for the individual regression coefficients are particularly significant, then we have multicollinearity. We can also check for linear relations between pairs of variables by looking at the values of the simple correlation coefficients between the Xs. A simple rule of thumb is that multicollinearity is likely to be important if $r_{X_i X_j} > R$ where $i \neq j$ and $R = \sqrt{(R^2)}$.

★ We can also check for linear relations between more than two
★ variables by examining the inverse of the matrix of simple
★ correlation coefficients between the Xs.

The model of import demand which we developed in the last chapter illustrates the problem of multicollinearity clearly. Let us take the version of the model where imports, M, depend upon GDP, Y, relative prices, P, the change in stocks, S, devaluation in 1967, D, and a time trend, t. The estimated equation is

$$M = 2{,}420 - 829P + 0.224S - 0.0464Y + 17.5t + 136D$$
$$ (650) \quad (500) \quad (0.155) \quad (0.0934) \quad (5.9) \quad (55)$$

We have a high R^2 of 0.971, but only the constant term, the time coefficient and the dummy variable are significantly different from zero at 5% and the income coefficient has a perverse sign. The reason is clear if we consider the correlation coefficient between Y and t which has the value 0.946.

The next question is how to avoid the problem once we have discovered its existence. There are a number of possible solutions. Firstly we might be able to specify a relationship between the collinear variables and hence estimate this relationship separately, thus avoiding the necessity of including them all in the main equation. Secondly it may be possible to obtain information about the values of the regression coefficients from elsewhere, for example from cross-section data. Lastly, it may be possible to obtain more observations on the variables, although this may not necessarily solve the problem and could even make it worse. A common solution which is used is merely to drop a variable from the analysis. Various criteria for choice are advanced for choosing the variable to drop, but the only proper criterion is the specification of the model. In our import example we have no real theoretical justification for including the time trend in the model, so we would omit this variable. If we had used some criterion related to obtaining the highest R^2 we would have omitted Y, which would have removed much of the economic content of the model. The reader is warned strongly against succumbing to the temptation of trying to obtain the highest possible R^2 without taking other factors into consideration.

8.6 Single-Equation Problems: Autocorrelation

In our original least-squares model we assumed that the errors in the model were independent, it is, however, often the case in economic time-series models that this assumption is not valid. Thus if our model is of the form

$$Y_t = b_1 X_{t1} + b_2 X_{t2} + \ldots + b_k X_{tk} + e_t \tag{20}$$

$E[e_i e_j] \neq 0$ when $i \neq j$. A simple example of this is that the errors between successive time periods are correlated

$$e_t = a e_{t-1} + u_t \tag{21}$$

where $E[u] = 0$ and $E[u_i u_j] = \sigma_u^2$ when $i = j$ and 0 if $i \neq j$. A series which is correlated with itself, lagged any number of periods, i.e. $r_{e_i e_{i-s}} \neq 0$ ($s \neq 0$), is said to be autocorrelated; thus the residuals in (20) have a first-order autoregressive relationship as shown in (21); other orders and forms of autocorrelation can and do exist as well. We would observe the situation shown by (20) and (21) if a producer is trying to supply a market according to the form of (4). However, information is not perfect; he does not know the prevailing price exactly but he has to sell in the market all that he has produced. We could assume for example that he is selling a perishable commodity. Thus in every period there is a u_t, but the producer tries to learn by his mistakes, and if he oversupplies in one period he cuts the amount he supplies in the next. If the producer overreacts a will be negative; and if he takes insufficient action, a will be positive but less than unity.

If we were to estimate a model which actually consisted of equations (20) and (21) as if only (20) existed, our estimators of the regression coefficients, $b_1 \ldots b_k$, would not be biased. However, we would underestimate both σ^2 and σ_b^2, and our underestimate would be the more extreme the farther a differs from zero. Thus the danger is that we shall draw incorrect inferences concerning the values of the parameters of the model. In particular we will tend to reject the null hypothesis that $b_i = 0$ when it is true because we have underestimated $\sigma_{\hat{b}_i}$. In the second place we shall also tend to reject the null hypothesis that $b_1 = b_2 = \ldots = b_k = 0$, because σ^2 is underestimated. It is thus important both to use a more efficient estimation method than ordinary least squares if we have autocorrelation and also to detect whether autocorrelation may exist in any particular case.

Let us consider this last point first. We may suspect that in fact we have (21) as well as (20) but wish to test the null hypothesis that $a = 0$. We can do this by calculating a statistic called *Durbin—Watson d* (d or *DW* for short).

$$DW = \frac{\sum\limits_{i=2}^{N} (\hat{e}_i - \hat{e}_{i-1})^2}{\sum\limits_{i=1}^{N} \hat{e}_i^2} \tag{22}$$

where the $\hat{e}_i$ are the residuals calculated from the ordinary least-squares estimate of (20) and N is the number of observations. If $a = 0$ DW will tend to the value 2. Using the fact that

$$\sum_{i=1}^{N-1} \hat{e}_i^2 \doteq \sum_{i=2}^{N} \hat{e}_i^2 \doteq \sum_{i=1}^{N} \hat{e}_i^2$$

we can expand (22) as follows:

$$\frac{\sum\limits_{i=2}^{N} (\hat{e}_i - \hat{e}_{i-1})^2}{\sum\limits_{i=1}^{N} \hat{e}_i^2} = \frac{\sum\limits_{i=2}^{N} \hat{e}_i^2 - 2\sum\limits_{i=2}^{N} \hat{e}_i \hat{e}_{i-1} + \sum\limits_{i=1}^{N-1} \hat{e}_i^2}{\sum\limits_{i=1}^{N} \hat{e}_i^2}$$

$$\doteq 1 - \frac{2\sum\limits_{i=2}^{N} \hat{e}_i \hat{e}_{i-1}}{\sum\limits_{i=1}^{N} \hat{e}_i^2} + 1$$

$$\doteq 2 - 2r_a = 2(1 - r_a) \tag{23}$$

where r_a is the correlation coefficient between $\hat{e}_i$ and $\hat{e}_{i-1}$. (Recalling that $\bar{\hat{e}} = 0$ and hence

$$\sum_{i=2}^{N} \hat{e}_i \hat{e}_{i-1} \doteq N\mathrm{Cov}(\hat{e}_i \hat{e}_{i-1})$$

We can see from (23) that if $r_a = 0$, $DW = 2$ and if $r_a = 1$, $DW = 0$. Also if we have negative autocorrelation ($a < 0$) $r_a = -1$ and DW will tend to 4.

The drawback of this statistic is that the $\hat{e}_i$ depend on the particular values of X. Hence its distribution is not exact. For any given significance level we can give a value, d_L, below which we will reject the null hypothesis of no positive autocorrelation and a second, higher, value, d_U, above which we will accept the null hypothesis. If on the other hand the observed value of DW lies between d_L and d_U the result is inconclusive and we can neither accept nor reject the null hypothesis. (Since the distribution is symmetric if our null hypothesis is that there is no negative autocorrelation we can perform precisely the same test on $4 - DW$.) In Appendix Table A3.5 we have set out the values of d_L and d_U for the 5% and 1% levels of significance tabulated by the number of observations, N, and the number of variables on the right-hand side of the equation (excluding the constant), K. As N increases, d_L and d_U move closer together and both move nearer to 2 as is shown in Figure 8.6; and as K rises, for any N, d_L decreases and d_U rises.

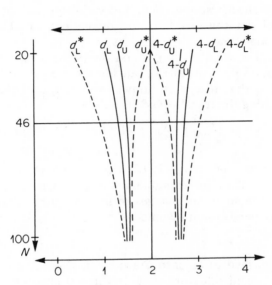

Figure 8.6 Approximate bounds of DW at the 5% level of significance. Solid lines represent $K = 1$ and dotted lines and asterisks $K = 5$

For example if $N = 46$, $d_L = 1.48$ and $d_U = 1.57$ when $K = 1$, and $d_L = 1.29$ and $d_U = 1.78$ when $K = 5$. If you look back to p. 137 you will recall that our import demand model had 46 observations and if we take the equation

$$M = b_0 + b_1 Y + b_2 PR + b_3 Q_1 + b_4 Q_2 + b_5 Q_3 + e$$

where $K = 5$ (all variables defined as before, p. 148) the estimates were

$$M = -850 + 0.406Y - 536PR + 254Q_1 + 161Q_2 + 166Q_3$$
$$(503) \quad (0.023) \quad (391) \quad (36) \quad (35) \quad (36)$$

and

$$DW = 0.675$$

Thus we reject the null hypothesis of no positive autocorrelation at the 5% level as $0.675 < 1.29$. The reader is strongly warned that the production of a regression equation with a high value for R^2 may be of little meaning if DW is significantly different from zero — unfortunately this situation occurs all too frequently in economic literature.

We must avoid falling into the trap of testing hypotheses concerning autocorrelation purely because a computer program prints out values of the Durbin–Watson statistic. It is usually only in the case of time-series data that we can attach any meaning to the autocorrelation of the residuals. Even if there is no autocorrelation we expect 2½% of the values of DW to be less than d_L and 2½% to be greater than $4 - d_L$ purely as a result of random sampling variation. We should also note that DW is biased towards 2 if the dependent variable, lagged one period, is included among the regressors.

The techniques of estimating a and providing a satisfactory estimation of the coefficients in (20) lie outside the scope of this book, but it is worth noting that if a were known we could incorporate (20) and (21) by estimating the transformed equation

$$Y_i - aY_{i-1} = b_1(X_{i1} - aX_{i-1,1}) + b_2(X_{i2} - aX_{i-1,2}) + \ldots + u_i$$

$$(24)$$

In the extreme case where $a = 1$, (24) in fact is merely an equation in first differences. Thus for example we might suggest that changes in consumption depend upon changes in income rather than purely that consumption depends upon income.

8.7 Conclusions

We have now pushed as far into the subject of economic statistics as is reasonable for an introductory text. In this last chapter we have shown that our treatment can only hope to be introductory as the complexity of economic systems requires complex methods for its estimation. However, elementary concepts do have an important role to play, firstly because some economic problems do lend themselves readily to straightforward treatment, and secondly because simple methods can be powerful analytic tools and can lay bare the consequences of the assumptions of economic theory.

The problems we have introduced in this chapter: identification, simultaneity, collinearity and relationships between disturbances and observations in different time periods, form the core of further study. However, our models have been linear or linear in simple transformations, and we have given little consideration to the possibility of non-linearities. We have also not considered models of probability processes over time nor models involving inequalities and constraints. These and other considerations can be examined with the help of economic statistics.

As our last comment let us recall the role of economic statistics, that we set out in the Introduction, in the light of the experience which has been accumulated in the intervening chapters. It is not meaningful to undertake a statistical examination of economic phenomena without the prior derivation of some economic theory concerning these phenomena. We must be able to define our variables before we can examine them and we must be able to suggest that populations of variates have certain sorts of parameters before we can estimate them. In general, our analysis has to begin with the specification of economic models and hypotheses concerning them. It is only then that we can estimate the parameters of these models and test the relevant hypotheses. We have seen that our standard linear model requires many strong assumptions to be made for it to be an appropriate estimation method. We have also considered the distributional properties that our variables must have for our testing procedures to be valid. It is not legitimate to assume that economic reality just happens to conform to any particular set of properties purely because these suit the statistical techniques that we have at our command. Obviously the statistical methodology we have chosen to develop has been selected because it does in fact rest upon a set of assumptions which do conform adequately to economic reality in many cases. However, this last chapter has shown that we must look at each individual case with care.

The value of economic statistics is not just that we can estimate the parameters of economic models and test hypotheses about them, but that we can use this information to make statements about situations other than those which we have observed. In particular we can make predictions or forecasts about the likely consequences of changes in economic variables. From the point of view of the government we can suggest the sorts of effects that will result from changes in variables which governments can control, such as taxation and expenditure. From the point of view of the firm we can suggest the sorts of effects that may follow from changes in the variables which it controls; size of the labour force, advertising expenditure and so on. Without actually having to wait for events to occur we can simulate the behaviour of our models over time, allowing for both changes in the variables which the policy maker can control and for random or specific fluctuations in the rest of the system which is outside his control. It is thus possible to examine how one might achieve a policy objective, whether at a macro- or micro-economic level, by more than a process of trial and error. Economic statistics is a powerful tool and it has been the purpose of this book to introduce the student to its use and to encourage further study.

We must, on the other hand, remember that, however powerful the technique we are using, the quality of the information it provides cannot be greater than the quality of both the theory and the empirical observation on which it is based. We cannot expect to make forecasts which are accurate to within 1% on the basis of data which are only accurate to within 10%. We can allow for such inaccuracies if we know something about their properties but we cannot eliminate them from our analysis. We have, therefore, striven throughout the book to provide examples and illustrations from real data so that the drawbacks imposed on our methodology from that source are made clear. It is part of the economic statistician's task to try to provide data of a quality sufficient for the validity of his analysis whether the data be drawn from published sources or from direct observation.

References and Suggested Reading

1. *Monthly Digest of Statistics*, H.M.S.O., monthly.

This is the main source of monthly data for the U.K. and the place where most summary information, the Retail Price Index, the Index of Industrial Production, etc., is to be found.

2. Further reading. There are many books on Econometrics of which that by Johnston cited at the end of Chapter 6 is one of the best.

C. F. Christ, *Econometric Models and Methods*, John Wiley, New York, 1966

has a long introductory section explaining what econometrics is trying to do, and incorporates a good exposition of the identification problem.

There are books such as

A. A. Walters, *An Introduction to Econometrics*, Macmillan, London, 1968

which attempt to examine the subject with a minimum of mathematics. Our personal recommendation is that books which try to avoid using linear algebra are actually more confusing, and that the reader should persevere.

Clarity will also be aided by the use of rather more applied books, for example

J. S. Cramer, *Empirical Econometrics*, North Holland, Amsterdam, 1969.

The best incentive will, however, come from the reading of the vast number of economics books which use these techniques and ideas.

Questions

1 Using the information given below, comment on the statistical difficulties shown by the following equation

$$C_t = 2.36 - 2.56\, C_{t-1} - 2.84\, S_{t-1} + 3.25\, Y_{t-1}, \qquad R^2 = 0.91$$

$$(0.96)\ (2.55) \qquad (2.60) \qquad (2.56) \qquad F = 23.59$$

where C_t is consumers' expenditure in period t in £000 mn,

S_t is personal saving in period t in £000 mn (subject to slight measurement error),

Y_t is total personal disposable income in period t in £000 mn,

t runs from 1 to 12 and indicates quarterly data from 1971 III—1974 II.

Show the relationship between the regressor variables and explain why it has occurred.

Criticize the model as an explanation of consumers' expenditure. Suggest and estimate a more realistic specification.

The matrix of sums of squares and cross-products is as follows — all variables measured in deviations from their means.

	C_t	C_{t-1}	S_{t-1}	Y_{t-1}	Y_t
C_t	903.230	896.262	96.670	992.585	1,001.151
C_{t-1}	896.262	889.585	95.886	985.124	993.449
S_{t-1}	96.670	95.886	10.615	106.466	108.105
Y_{t-1}	992.585	985.124	106.466	1,091.209	1,100.210
Y_t	1,001.151	993.449	108.105	1,100.210	1,109.736

The data are drawn from *Economic Trends*, Oct. 1974.

2 The traditional macro-economic consumption function

$$c = a + by + e \tag{1}$$

where c is consumers' expenditure and y is GDP, is only one of two macro-economic relationships linking c and y. We can also recall the national accounting identity that

$$y = c + p + v + (x - m) \tag{2}$$

where p is public authorities' current expenditure, v is gross domestic capital formation, x is exports of goods and services and m is imports of goods and services.

Estimate (1) taking the existence of (2) into account using the data given below for the United Kingdom over the years 1962—72.

What effect does the existence of (2) have on the parameters of (1)? Is this effect important?

United Kingdom, £'000 mn 1970

Year	Consumers' expenditure	GDP at market prices	Public authorities' current expenditure	Gross domestic capital formation	Balance of trade $(x - m)$
1962	26.3	40.3	7.9	6.3	−0.2
1963	27.4	41.9	8.0	6.7	−0.2
1964	28.3	44.3	8.1	8.5	−0.6
1965	28.7	45.3	8.4	8.4	−0.2
1966	29.3	46.2	8.6	8.4	−0.1
1967	29.8	47.3	9.1	8.9	−0.5
1968	30.6	48.9	9.2	9.4	−0.3
1969	30.6	49.4	9.0	9.4	0.4
1970	31.4	50.4	9.1	9.5	0.4
1971	32.2	51.6	9.4	9.3	0.7
1972	34.1	52.7	9.8	9.0	−0.2

Source: *National Income and Expenditure*, 1973

3   Test the hypothesis that the number of injured persons absent from work owing to sickness and invalidity is a linear function of (a) the weather — as indicated by mean daily air temperature at sea level — (b) the size of the labour force.

Also examine the hypothesis that there should be a second equation in the model of the form of a first-order autoregressive process in the residuals from the first equation.

Great Britain

Period (1972)	Injured persons absent from work owing to sickness and invalidity (millions)	Employees in employment: production industries (millions)	Mean daily air temperature at sea level England and Wales (degrees centigrade)
January	1.023	10.142	4.5
February	1.140	10.090	5.0
March	1.040	10.041	7.0
April	0.943	10.052	8.8
May	0.938	10.042	11.1
June	0.919	10.029	12.4
July	0.912	10.055	15.7
August	0.892	10.074	15.6
September	0.907	10.052	12.4
October	0.962	10.063	11.2
November	1.017	10.097	6.9
December	1.103	10.076	6.6

Source: *Monthly Digest of Statistics*

4   If we use the data in the table below we can estimate the following investment function

$$I_t = -1,544 + 0.384Y_t - 32.40R_t, \quad R^2 = 0.945$$
$$(231) \ (0.034) \quad (17.18) \qquad d = 0.663$$

where I_t is gross domestic fixed investment (1963 prices),

Y_t is gross domestic product (1963 prices),

R_t is rate of interest (yield on 2½% consols),

t indicates the time period — 40 quarters 1962 I—1971 IV.

Comment on the plausibility of the model and the estimates of it.

Test the hypothesis that there is positive first-order autocorrelation of the residuals.

Using a simple transformation re-estimate the model so as to reduce the problem of autocorrelation — comment on your results.

United Kingdom

t	I £mn	Y £mn	R %	t	I £mn	Y £mn	R %
1	1,416	8,493	5.80	21	2,246	10,313	7.13
2	1,622	8,920	5.54	22	2,217	10,276	7.30
3	1,674	8,919	5.39	23	2,225	10,489	7.47
4	1,739	9,122	5.62	24	2,263	10,557	7.69
5	1,804	9,174	5.91	25	2,279	10,501	8.42
6	1,855	9,438	6.04	26	2,235	10,471	9.11
7	1,911	9,374	6.04	27	2,280	10,522	9.15
8	1,934	9,520	6.13	28	2,248	10,662	8.87
9	1,959	9,555	6.30	29	2,236	10,514	8.60
10	1,921	9,539	6.55	30	2,317	10,708	9.23
11	1,921	9,643	6.48	31	2,318	10,736	9.28
12	2,006	9,730	6.34	32	2,352	10,830	9.59
13	1,974	9,712	6.53	33	2,299	10,686	9.38
14	1,964	9,757	6.78	34	2,332	10,838	9.24
15	2,018	9,841	7.11	35	2,318	11,111	9.05
16	2,027	9,911	6.78	36	2,320	11,033	8.93
17	2,091	10,016	6.47	37	2,366	10,921	8.45
18	2,184	10,085	6.51	38	2,341	11,077	8.98
19	2,163	10,070	6.82	39	2,321	11,067	9.46
20	2,126	9,950	7.01	40	2,354	11,279	9.63

Source: *Economic Trends*

5   The following model has been used to estimate the determinants of inflation. (This model is presented by R. G. Lipsey and M. Parkin in Parkin and Sumner (Eds.), *Incomes Policy and Inflation*.)

$$\dot{p} = a_0 + a_1\dot{w} + a_2\dot{m}_{-1} + a_3\dot{q}$$

$$\dot{w} = b_0 + b_1U + b_2\dot{p} + b_3\dot{N}$$

where p refers to the retail price index,

w refers to the index of wage rates (all industries),

m refers to import prices,

q refers to output per head,
U refers to unemployment (%),
N refers to per cent of labour force belonging to a trade union,
· denotes proportionate rate of change,
-1 denotes a one period lag,
the a_i and b_i denote parameters.

What sort of values would you expect to observe for the coefficients in the equations?

The equations were estimated by ordinary least squares using quarterly data over the period 1948—68 for the United Kingdom, and the following results were obtained:

$$\dot{p} = 1.374 + 0.562\dot{w} + 0.085\dot{m}_{-1} - 0.145\dot{q}, \quad R^2 = 0.697$$
$$\quad (2.51) \quad (5.33) \quad\quad (4.60) \quad\quad\quad (3.48) \quad\quad DW = 0.946$$

$$\dot{w} = 4.147 - 0.891U + 0.482\dot{p} + 3.315\dot{N}, \quad R^2 = 0.616$$
$$\quad (4.26) \quad (1.77) \quad\quad (5.76) \quad\quad (2.09) \quad\quad DW = 0.742$$

N.B. Numbers in parentheses are t-values.

Do these results seem plausible? Comment on them in the light of your expectations.

Does it look legitimate to use ordinary least squares as an estimation method? Do the statistics which have been calculated indicate the presence of any problems? If a better estimation method were used, what sort of effect would this be likely to have on the estimates and the results of tests of hypotheses concerning them?

Answers
1   It is immediately clear that the model is not only badly specified, but also mal-estimated. We know that in any one time period t

$$Y_t \equiv C_t + S_t$$

Thus the regressor variables are clearly linearly related. In fact were it not for the measurement error in S it would have been impossible to estimate the equation at all. We can see the usual signs of multicollinearity: the standard errors are relatively large given the value of R^2. We can also see that the values of the coefficients make no real sense; they are too large in absolute terms and the model is unstable. (It will generate increasing fluctuations over time.)

The correlation matrix between regressors is

	C_{t-1}	S_{t-1}
Y_{t-1}	0.952	0.813
C_{t-1}		0.596

and the multiple correlation coefficient between C_{t-1} and S_{t-1} and Y_{t-1} is 0.998. Thus while S_{t-1} is fairly independent of both Y_{t-1} and C_{t-1} individually it is highly related to them jointly. In this case the multi-collinearity is evident through the correlation between C_{t-1} and Y_{t-1} but sometimes we need to consider higher orders of correlation to determine the source of the problem.

An alternative more plausible specification might be the permanent income hypothesis of the form

$$C_t = a_0 + a_1 C_{t-1} + a_2 Y_t + e$$

Using the information in the table we can estimate this new model. We require the matrix of sums of squares and cross-products between C_t, C_{t-1} and Y_t which is as follows:

	C_t	C_{t-1}	Y_t
C_t	903.230	896.262	1,001.051
C_{t-1}	896.262	889.585	993.449
Y_t	1,001.051	993.449	1,109.736

The estimated equation is thus

$$C_t = 1.69 + 0.33 C_{t-1} + 0.44 Y_t, \quad R^2 = 0.889$$
$$\quad (0.90)\ (0.29) \qquad (0.24) \qquad F\ = 31.99$$

This is a rather more plausible equation, although it is still clear from the relationship between R^2 and the t statistics that the multi-collinearity has not been fully overcome.

2   We could estimate the parameters of (1) in two ways -- firstly by estimating the parameters of the reduced form of the model by ordinary least squares and substituting back to obtain the structural coefficients (Indirect Least Squares) or secondly by using Two-Stage Least squares.

The first point to notice is that the three exogenous variables in the model, p, v and $(x - m)$, have an additive and deterministic effect on y. We can, therefore, aggregate them, $z = p + v + (x - m)$ and use z as a single exogenous variable in the model.

The reduced form is easily obtained. Substituting from (1) into (2)

$$y = a + by + e + z$$

$$y - by = a + z + e$$

$$y = \frac{a}{1-b} + \frac{z}{1-b} + \frac{e}{1-b} \tag{3}$$

Substituting in (1) from (2)

$$c = a + b(c + z) + e$$

$$c - bc = a + bz + e$$

$$c = \frac{a}{1-b} + \frac{b}{1-b} z + \frac{e}{1-b} \qquad (4)$$

Firstly, using Indirect Least Squares, we estimate (3) and (4) by ordinary least squares

$$y = 10.94 + 2.099z, \qquad R^2 = 0.939$$
$$(3.09) \quad (0.178)$$
$$c = 10.94 + 1.099z, \qquad R^2 = 0.808$$
$$(3.09) \quad (0.178)$$

hence

$$\frac{a}{1-b} = 10.94$$

$$\frac{b}{1-b} = 1.099$$

$$\frac{1}{1-b} = 2.099$$

Solving for b

$$\frac{b}{1-b} \bigg/ \frac{1}{1-b} = b = \frac{1.099}{2.099} = 0.524$$

Solving for a

$$a = (10.94)(1 - b) = (10.94)(0.476) = 5.21$$

Thus by ILS

$$c = 5.21 + 0.524y$$

Using Two-Stage Least Squares we must first of all obtain an estimate of the values of the endogenous variable, y, on the right-hand side of (1) by regressing it on the exogenous variable, z. Thus the first stage is to estimate

$$y = c + dz + u \qquad (5)$$

and then calculate

$$\hat{y} = \hat{c} + \hat{d}z \qquad (6)$$

The second stage is then completed by the regression of the dependent variable c on $\hat{y}$

$$c = a + b\hat{y} + e*$$

$$= a + b\hat{y} + bu + e$$

Proceeding to the first stage, the ordinary least-squares estimate of (5) is

$$y = 10.94 + 2.099z$$

From this we obtain $\hat{y}$

£'000mn	Year	1962	1963	1964	1965	1966	1967
	$\hat{y}$	40.326	41.376	44.524	45.783	46.413	47.673

£'000 mn	Year	1968	1969	1970	1971	1972
	$\hat{y}$	49.352	50.401	50.821	51.661	49.981

Then estimating (7)

$$c = 5.21 + 0.524\hat{y}$$

$$(4.01)\ (0.085)$$

Thus our 2SLS and ILS estimates of (1) are identical.

If we compare this result with that obtained by ordinary least squares

$$c = 3.845 + 0.553y, \quad R^2 = 0.959$$

$$(1.796)\ (0.038) \quad F = 212$$

We can see that the differences between the two equations are small. The OLS estimates do exhibit the expected difference from the 2SLS estimates. If we use OLS $\hat{b}$ is biased upwards and $\hat{a}$ downwards. In this case the OLS estimate of $\hat{b}$ is larger than the unbiased estimate from 2SLS and the estimate of $\hat{a}$ is lower.

3 Let us assume that injured persons, I, are a linear function of employees in employment, E, and temperature, T.

$$I = a_0 + a_1 E + a_2 T + e$$

We can estimate this by ordinary least squares

$$I = 1.965 - 0.0797E - 0.0184T, \quad R^2 = 0.742$$

$$(5.437)\ (0.537) \quad (0.0043)$$

It appears that the labour force has no real effect, and what effect it does have is negative. Temperature, on the other hand, has the effect we might expect — the higher the temperature the smaller the number of persons injured.

If we wish to test the hypothesis that we have first-order autocorrelation we must calculate the Durbin—Watson d statistic, where

$$d = \frac{\sum\limits_{i=2}^{N} (\hat{e}_i - \hat{e}_{i-1})^2}{\sum\limits_{i=1}^{N} \hat{e}_i^2}$$

We must therefore begin by calculating

$$\hat{e}_i = I_i - \hat{I}_i = I_i - \hat{a}_0 - \hat{a}_1 E_i - \hat{a}_2 T_i$$

for all i.

i	I_i	$\hat{I}_i$	$\hat{e}_i$	$(\hat{e}_i^2)(10^4)$	$\hat{e}_i - \hat{e}_{i-1}$	$(\hat{e}_i - \hat{e}_{i-1})^2(10^4)$
1	1.023	1.074	−0.051	26.01	—	
2	1.140	1.069	0.071	50.41	0.122	148.84
3	1.040	1.036	0.004	0.16	−0.067	44.89
4	0.943	1.002	−0.059	34.81	−0.063	39.69
5	0.938	0.960	−0.022	4.84	0.037	13.69
6	0.919	0.938	−0.019	3.61	0.003	0.09
7	0.912	0.875	0.037	13.69	0.056	31.36
8	0.892	0.875	0.017	2.89	−0.020	4.0
9	0.907	0.936	−0.029	8.41	−0.046	21.16
10	0.962	0.957	0.005	0.25	0.034	11.56
11	1.017	1.033	−0.016	2.56	−0.021	4.41
12	1.103	1.041	0.062	38.44	0.078	60.84
Σ				186.08		380.53

$d = 380.53/186.08 = 2.04$

As d_U at 5% where $N = 12$ and $K = 2$ is 1.54 we accept the null hypothesis of no autocorrelation.

4　　The investment function specified here is of a simple Keynsian form. It takes no account of lags in the system or of expectations. Empirical research has also shown that the rate of interest is a rather weak determining variable. However, within these limits the estimated equation looks quite plausible. There is a strong relationship between GDP and investment and a weak negative relationship between the rate of interest and investment. The signs of the coefficients are at least those which are expected. The overall level of explanation indicated by R^2 is high.

When we come to look at the hypothesis of autocorrelation we can see immediately that with a value of Durbin—Watson d of 0.663 we must reject the null hypothesis of no positive autocorrelation since d_L at 5%, $N = 40$, $K = 2$ is 1.39.

A simple solution to this problem is to take first differences of the model — in other words to assume that the autoregressive equation is of the form

$$e_t = e_{t-1} + u_t$$

The suggestion that it is changes in the variables which are related is not totally unreasonable. It is certainly no worse than any other arbitrarily chosen autoregressive relation.

The first stage must therefore be to calculate first differences and then to re-estimate the model. The transformed data are shown in the table and the estimated equation becomes

$$I_t - I_{t-1} = 10.76 + 0.2317(Y_t - Y_{t-1}) - 30.24(R_t - R_{t-1})$$
$$(7.95)\ (0.0508) \qquad\qquad (23.34)$$

$$R^2 = 0.386,\ d = 2.08$$

Although the signs of the coefficients remain the same and their values are fairly stable (the constant, of course, excepted), standard errors are increased and the overall fit of the estimated equation falls. This is the result we expect as the OLS estimators of the coefficients are unbiased, but OLS underestimates the variance of the estimators. The value of d so close to 2 tends to show that our transformation has been successful.

t	$I_t - I_{t-1}$	$Y_t - Y_{t-1}$	$R_t - R_{t-1}$	t	$I_t - I_{t-1}$	$Y_t - Y_{t-1}$	$R_t - R_{t-1}$
1				21	120	363	0.12
2	206	427	−0.26	22	−29	−37	0.17
3	52	−1	−0.15	23	8	213	0.17
4	65	203	0.23	24	38	68	0.22
5	65	52	0.29	25	16	−56	0.73
6	51	264	0.13	26	−44	−30	0.69
7	56	−64	0.00	27	45	51	0.04
8	23	146	0.09	28	−32	140	−0.28
9	25	35	0.17	29	−12	−148	−0.27
10	−38	−16	0.25	30	81	194	0.63
11	0	104	−0.07	31	1	28	0.05
12	85	87	−0.14	32	34	94	0.31
13	−32	−18	0.19	33	−53	−144	−0.21
14	−10	45	0.25	34	33	152	−0.14
15	54	84	0.33	35	−14	273	−0.19
16	9	70	−0.33	36	2	−78	−0.12
17	64	105	−0.31	37	46	−112	−0.48
18	93	69	0.04	38	−25	156	0.53
19	−21	−15	0.31	39	−20	−10	0.48
20	−37	−120	0.19	40	33	212	0.17

5 The model indicates a two-way relationship between the proportional rates of change of wages and prices. For this relationship to be stable, $-1 < a_1 < 1$ and $-1 < b_2 < 1$. However, we would expect a positive relationship between the two variables (if wages increase faster,

prices increase faster, and vice versa). Thus we can say firstly

$$0 < a_1 < 1$$

and

$$0 < b_2 < 1$$

Secondly, since import prices will form part of retail prices we would expect

$$0 < a_2 < 1$$

but that the effect would be nearer zero than unity as they do not form a major part.

If we take the unemployment rate as being some sort of measure of the excess supply of labour, then we would expect that the greater the excess supply the slower the rate of increase in wages, thus

$$b_1 < 0$$

Fourthly, if centralized bargaining is more effective and unions are good negotiators, we would anticipate that

$$b_3 > 0$$

As far as productivity is concerned, for any given values of the other variables we would expect that if output per head is rising faster there will be less pressure on prices, so

$$a_3 < 0$$

This leaves us with the constant terms. It is very difficult to think of any specific reason why they should be positive rather than negative, or even why they should be non-zero. We, therefore, have no explicit expectation, we merely think that there is likely to be a constant element in both relationships.

When we look at the estimated equations it is apparent that all the coefficients have the expected signs. All are significantly different from zero at the 5% level with the exception of the unemployment term in the wage equation. This latter is perhaps not surprising in view of much recent scepticism about the existence of a Phillips Curve in the U.K. The results thus seem very satisfactory.

However, when we consider the suitability of ordinary least squares we can see immediately that we have a simultaneous equations problem. Both equations have an endogenous variable on the right-hand side. We would thus expect that the coefficients estimated by OLS will be subject to simultaneous equation bias.

The second striking feature is the values of Durbin–Watson d. In both cases we would reject the null hypothesis of no positive first-order autocorrelation. Our standard assumption about the distribution of the disturbances would thus appear to be false. If we believe that the true model includes a first-order autoregressive process in the residuals, then our estimation method should take account of this.

Thus we are suggesting that a more satisfactory estimation method would take account of both simultaneity and autocorrelation of the residuals. The result of using this method will be firstly that the estimated values of the coefficients are likely to change as a result of the resolution of the problem of simultaneity (the greater the problem the greater the likely change). In the second place the allowance for autocorrelation will result in a general lowering of significance levels, and hence we may very well accept null hypotheses that were rejected when OLS was used. When a suitable method is used the estimated equations run

$$\dot{p} = 0.903 + 0.731\dot{w} + 0.052\dot{m}_{-1} - 0.230\dot{q}, \hat{c} = 0.528, s^2 = 0.144$$
$$(3.070)\,(0.531)\quad(0.048)\qquad(0.244)\qquad(0.119)$$
$$\dot{w} = 2.957 - 0.674U + 0.728\dot{p} - 1.193\dot{N}, \hat{c} = 0.641, s^2 = 1.044$$
$$(3.397)\,(1.556)\quad(0.336)\quad(7.616)\qquad(0.096)$$

N.B. Numbers in parentheses are standard errors. $\hat{c}$ is the estimated value of c in the first-order autoregressive relationship $e_t = ce_{t-1} + u_t$. s^2 is the estimate of σ_u^2.

★ (The method used is known as Autoregressive Instrumental Vari-
★ ables and was estimated by L. G. Godfrey in *Incomes Policy and*
★ *Inflation*, Eds. Parkin and Sumner, 1973, Manchester University
★ Press.)

It is readily seen that the results are drastically different from those shown in the question; only the price coefficient in the wage equation is significantly different from zero at 5%.

Appendix 1 Index Numbers

Index numbers form a very important weapon in the economic statistician's power. Their consideration has been relegated to this Appendix firstly because we cannot hope to cover the whole subject in depth, and secondly because the rest of the book can be read satisfactorily without them.

A1.1 Comparisons of a Single Variable over Space and Time
There are two main uses of index numbers, firstly to facilitate the comparison of variables over time, space and different units and secondly in the formation of a single measure to represent a number of different variables. We can consider index numbers most easily by taking these two points in order. An index number is a ratio of two values of a variable or group of variables expressed in the same units, where our purpose is to compare the first value to the second. Thus in the simplest case where V_1 is the first value and V_0 the second, an index I can be formed from the ratio V_1/V_0. Conventionally the ratio is multiplied by 100 so that $I = (V_1/V_0)(100)$. We have shown a simple example of the use of index numbers in the second column of Table A1.1. Our aim is to examine how GNP in the U.K. increased over the 1950s. We must begin by deciding how we are going to make the comparison, and in this case we have decided that we are going to compare GNP in each year with GNP in 1951. GNP in 1951 is thus the 'basis' of the comparison, our V_0, and 1951 is referred to as the '*base*' year of the index. We could choose any year as the base, it does not have to be the first.

Using the values of GNP given in the first column of Table A1.1 we can now derive the second; the index number for 1952 is, for example, $I_{1952} = (V_{1952}/V_{1951})(100) = (16{,}455/16{,}561)(100) = 99.4$. GNP in 1952 was thus 0.6%, $(100 - 99.4)$, lower than it was in 1951. Looking

Table A1.1
Gross National Product (1954 prices)

Year	U.K. £mn	U.K. 1951 = 100	France mn NF	France 1951 = 100
1951	16,561	100.0	143,730	100.0
1952	16,455	99.4	147,350	102.5
1953	17,167	103.7	151,840	105.6
1954	17,903	108.1	159,190	110.8
1955	18,404	111.1	168,450	117.2
1956	18,894	114.1	176,910	123.1
1957	19,225	116.1	187,500	130.5
1958	19,427	117.3	190,790	132.7
1959	20,113	121.4	195,310	135.9
1960	20,966	126.6	207,920	144.7

Source: OECD

further down the second column we can read off directly that GNP had increased by 17.3% by 1958 compared to 1951 and so on. This, therefore, enables us to make easy comparisons over time. It is also worth noting that the measure is unit-free so we can compare the increase in the U.K.'s GNP with that of France directly if we use the index numbers in the table. We can see that by 1958 French GNP has increased by 32.7% compared to its level in 1951.

Secondly we can compare different values, which are in the same units of measurement, observed in the same time period. For example, in Table A1.2 we have recorded the average price of butter imported into the United Kingdom during December 1972, and the average wholesale price of the domestic product. The second column in the table shows the index of butter prices with the U.K. price as the base. Again in this case we can see that the price of Danish butter was 29.7% higher than that of U.K. butter, the price of Polish butter was only 7.7% higher and so on.

Table A1.2
Average Price of bulk butter traded in December
1972 in U.K.

Country of origin	£/cwt	U.K. = 100
U.K.	18.49	100.0
Denmark	23.99	129.7
Netherlands	20.19	109.2
Irish Republic	23.43	126.7
France	20.60	111.4
Poland	19.91	107.7
New Zealand	20.26	109.6
Australia	21.22	114.8

Source: U.K. Overseas Trade Accounts, Commonwealth Secretariat

A1.2 Weighted Index Numbers

Thus far we have only considered the formation of index numbers from values of a single variable; however, they have a more important use with groups of variables. We commonly want measures of aggregate performance, the output of an industry, the price of food, the cost of living and so on. The simplest way to approach this would be to form an index from the sum of the aggregate,

$$I_j = \frac{\sum\limits_{k} V_{jk}}{\sum\limits_{k} V_{0k}}$$

where we are comparing situation j with the base 0 and the subscript k refers to the individual components of the group of values, prices, output, etc., which are to be aggregated. This method will only be appropriate if each of the k items are measured in the same units, we cannot, for example, add together tons of coal, gallons of oil, cubic feet of gas and kilowatts of electricity in order to form an index of the output of fuels. We can overcome this problem though by summing the individual ratios, (V_{jk}/V_{0k}), over k and averaging

$$I_j = \frac{\sum\limits_{k=1}^{n} (V_{jk}/V_{0k})}{n}$$

where n is the number of individual values. However, it is often not possible to arrive at these aggregate figures simply by adding together a number of components. Let us take the price of food as an example. There is no one price for food, and a useful aggregate would not be arrived at purely by taking an arithmetic average of the prices of all the different foodstuffs found in a shop or set of shops. It is very unlikely that we would consider the price of a water melon and the price of a bag of potatoes as equally important in determining the price of food in the U.K.

Let us begin by suggesting why we might be interested in the price of food. A simple suggestion might be that expenditure on food forms a sizable proportion of our total spending and we like to have an idea of how this expenditure is likely to change. We are concerned, therefore, with the price of the groups of foodstuffs which we normally buy. Since we have talked about 'expenditure' we also have an interest in the quantities we buy. However, in this instance we wish to separate the effects of changes in price and changes in quantity. The fact that we are interested in changes in prices rather than price levels means that we can use an index number and have no problems about units of measurement. We do not need to worry about the meaning of such concepts as 'so many pence per pound of food.' We can, therefore, list the products in which we are interested and then try to attach a degree of importance to each of them. Clearly there are many such methods of attaching importance, and we can express any such index number in the form

$$I_j = \frac{\sum_k (P_{jk}/P_{0k})w_k}{\sum_k w_k} (100) \tag{1}$$

where w_k is the weight, or importance attached to item k and P_{jk} refers to the price of item k in period j, period 0 being the base year against which we are making our comparison. We can see immediately that if we give our different foods an equal weight I_j is an ordinary arithmetic average of the individual price changes. I_j is computed for a single group of goods using equal weights of unity in Table A1.3. The prices of each individual product are shown in the first three columns of the table, labelled P_0, P_1 and P_2, and the ratios P_1/P_0 and P_2/P_0 are shown in the last two columns of the second part of the table, the index itself being shown in the first row of the last part of the table, $I_0 = 100$, $I_1 = 119$, $I_2 = 145$.

In practice we require some set of weights which is meaningful and easily obtainable. The most frequent solution is to make use of the quantities purchased of the goods. In Table A1.3 we have set out the prices and quantities purchased in each of three years. We can obviously use the quantities from any of these three years to derive a set of weights. Two of the most well-known types of index number, Paasche and Laspeyres, make use of the current year and base year quantities respectively for weighting. The Laspeyres index uses expenditure in the base year, $P_{0k}Q_{0k}$, as the weight; thus $w_k = P_{0k}Q_{0k}$ and

$$LI_j = \frac{\sum_k (P_{jk}/P_{0k})P_{0k}Q_{0k}}{\sum_k P_{0k}Q_{0k}} (100) \tag{2}$$

which we can simplify as

$$LI_j = \frac{\sum_k P_{jk}Q_{0k}}{\sum_k P_{0k}Q_{0k}} (100) \tag{3}$$

Equation (2) shows the Laspeyres index as an average of price relatives weighted by expenditure in the base year and equation (3) shows it as a ratio of the sum of the individual prices weighted by quantities purchased in the same year.

It is difficult to express the Paasche index in the same form as (2),

Table A1.3
Formation of Price Index Numbers

Item	Price at			Quantities purchased		
	March 1972 (P_0)	March 1973 (P_1)	March 1974 (P_2)	March 1972 (Q_0)	March 1973 (Q_1)	March 1974 (Q_2)
(1) Rump steak P = p/lb, Q = lb	69.6	87.0	91.4	2	1	1
(2) Bread, white P = p/loaf, Q = 14 oz loaves	6.2	6.6	9.5	4	5	3
(3) Potatoes, white P = p/lb, Q = lb	1.9	2.2	2.5	20	20	25
(4) Oranges P = p/lb, Q = lb	7.5	8.8	9.9	3	2	3
(5) Eggs, large P = p/doz, Q = doz	23.4	30.5	41.7	3	2	1

Source of price data: *Department of Employment Gazette*

	P_0Q_0	P_1Q_1	P_2Q_2	P_0Q_1	P_0Q_2	P_1Q_0	P_2Q_0	P_1/P_0	P_2/P_0
(1)	139.2	87.0	91.4	69.6	69.6	174.0	182.8	1.25	1.31
(2)	24.8	33.0	28.5	31.0	18.6	26.4	38.0	1.06	1.53
(3)	38.0	44.0	62.5	38.0	47.5	44.0	50.0	1.16	1.32
(4)	22.5	17.6	29.7	15.0	22.5	26.4	29.7	1.17	1.32
(5)	70.2	61.0	41.7	46.8	23.4	91.5	125.1	1.30	1.78
Total	294.7	242.6	253.8	200.4	181.6	362.3	425.6	5.94	7.26

	1972	1973	1974
Unit weights	100	$\dfrac{\Sigma P_1/P_0}{5}(100) = 119$	$\dfrac{\Sigma P_2/P_0}{5}(100) = 145$
Laspeyres	100	$\dfrac{\Sigma P_1 Q_0}{\Sigma P_0 Q_0}(100) = 123$	$\dfrac{\Sigma P_2 Q_0}{\Sigma P_0 Q_0}(100) = 144$
Paasche	100	$\dfrac{\Sigma P_1 Q_1}{\Sigma P_0 Q_1}(100) = 121$	$\dfrac{\Sigma P_2 Q_2}{\Sigma P_0 Q_2}(100) = 140$

but in terms of (3) the Paasche index, PI_j, uses the quantities from the current year, Q_{jk}, as weights, instead of the quantities in the base year, Q_{0k}. Thus

$$PI_j = \frac{\sum_k P_{jk} Q_{jk}}{\sum_k P_{0k} Q_{jk}} \, (100) \tag{4}$$

When we consider the results from Table A1.3 we can see that the three index numbers give different values for the level of prices in each of the two years 1 and 2 compared to year 0. Since their method of calculation differs this is not surprising, but it does show that the weights chosen are of importance. The particular choice depends on what we wish to compare. If we wish to know how the price of goods we used to buy in the base year has changed we would use a Laspeyres index and if we wished to know how the price of the collection of goods we bought in each period changed then we would use a Paasche index. The reader will observe that the Laspeyres index indicates a faster increase in prices than does the Paasche index. This is as a result of simple economic behaviour. We will always expect Paasche index numbers to lie below Laspeyres index numbers in times of rising prices and to lie above them in times of falling prices, because consumers will tend to substitute against these goods which have risen relatively fast in price (fallen relatively slowly). If the components of the index are not substitutes for each other then clearly we will not expect to observe such effects. In the example in Table A1.3 we can see from the last column (headed P_2/P_0) that bread and eggs have risen fastest in price, and from the columns Q_0 and Q_2 we can see that there has been a substitution away from these and towards the consumption of potatoes. However, since we have only illustrated the problem for a small group of foodstuffs we cannot expect this relationship to be particularly strong.

A1.3 Measurement at Constant Prices and the Deflation of a Series
In Table A1.1 and at a few other places in the book we have used the concept of measurement of a variable in the prices of a particular year. We can relate this concept to our consideration of index numbers. Clearly if any variable, say output, is measured in the prices of year j then for any other year i the variable has the value $P_j Q_i$ and not the value in current prices, $P_i Q_i$. To obtain the variable in the prices of a particular year, often referred to as 'constant prices', we must divide the values in current prices by a price index of the variable, as $P_j Q_i$ is not itself directly observable. This process of division by a price index is known as 'deflation', and the price index is called a 'deflator'. For example if we wished to deflate a series by a Paasche price index we

Table A1.4
General Index of Retail Prices. *16 January 1962 = 100*

	All items	Food	Alcoholic drink	Tobacco	Housing	Fuel and light	Durable household goods	Clothing and footwear	Transport and vehicles
Weights									
1971	1,000	250	65	59	119	60	61	87	136
1972	1,000	251	66	53	121	60	58	89	139
Index									
1971	153.4	155.6	152.7	138.5	172.6	160.9	135.4	132.2	147.2
1972 (Monthly averages)	164.3	169.4	159.0	139.5	190.7	173.4	140.5	141.8	155.9

Source: *Monthly Digest*

could calculate

$$P_i Q_i / PI_i = P_i Q_i \left/ \left(\frac{P_i Q_i}{P_0 Q_i} \right) \right. = P_0 Q_i$$

If the variable we are deflating is complex, such as GDP, then the price index will be a complex sum of prices. Output measured at constant prices can be made into a quantity index by dividing by the value of the base year, $I_i = P_0 Q_i / P_0 Q_0$, where quantities in each year are weighted by prices of the base year. Our concentration on price index numbers here is only for illustrative purposes; we can construct Paasche and Laspeyres quantity indexes, and weighted indexes of any economic variable.

A1.4 The Index of Retail Prices

This, therefore, provides an outline of the principle of index numbers; the reader should consult a more specific text if he wishes to carry the problem further (for example Merrill and Fox, Chapter 3). However, before leaving the subject it is worth noting that the retail price index published in the United Kingdom uses a more complex form of weighting which we have not considered. The weights vary over time and are updated every year to take account of changes in consumption patterns. They are an average of values derived from the previous three Family Expenditure Surveys. A continuous index with such a system of weights is known as a 'chained' index. An excerpt from the Index of Retail Prices is shown in Table A1.4.

References and Suggested Reading

1. W. C. Merrill and K. A. Fox, *Introduction to Economic Statistics*, John Wiley, New York, 1970, Chapter 3, provides a good exposition of the meaning and use of index numbers.
2. *Overseas Trade Accounts*, H.M.S.O., annually.
   Dairy Produce Review, Commonwealth Secretariat, London, occasionally.
   Department of Employment Gazette, H.M.S.O. monthly.

Appendix 2 Elements of Matrix Algebra

The purpose of this appendix is to provide a sufficient understanding of linear or matrix algebra to make it possible to generalize the principles of multiple regression. The initial driving force behind linear algebra was to find a convenient method of solving and manipulating a system of simultaneous linear equations. Our problem is to solve the normal equations which are set out on p. 139.

However, let us begin by explaining the concept of a matrix. A *matrix* is a two-dimensional array. For example, we can arrange the four numbers, 4, 2, 6 and 8 as a matrix

$$\begin{bmatrix} 4 & 2 \\ 6 & 8 \end{bmatrix}$$

which has two rows and two columns, usually referred to as a 2 by 2 or 2×2 matrix. Let us call this matrix A. We can, by using two subscripts, the first describing the row and the second the column, refer to every element in A. Hence

$$A = \begin{bmatrix} A_{11} & A_{12} \\ A_{21} & A_{22} \end{bmatrix} = \begin{bmatrix} 4 & 2 \\ 6 & 8 \end{bmatrix}$$

where $A_{11} = 4$, $A_{12} = 2$ and so on. The order of the elements in the matrix is thus important.

If we have two matrices A and B, $A = B$ only if $A_{ij} = B_{ij}$ for all i and j. This also tells us that A and B must each have the same number of rows and the same number of columns. If a matrix only consists of a single row or column then it is known as a '*vector*'. A well used convention to distinguish between the two is to refer to matrices by capital letters and column vectors by small letters. A row vector is denoted by a small letter followed by a dash. For example if we have a column vector

$$\mathbf{b} = \begin{bmatrix} 6 \\ 7 \\ 9 \end{bmatrix}$$

we could denote a row vector with the same elements

$$[6 \quad 7 \quad 9]$$

by $\mathbf{b}'$. The same sign is used for matrices and denotes a *transposed matrix*: if $\mathbf{B} = \mathbf{A}'$ then $B_{ij} = A_{ji}$ for all i and j. It follows, therefore, that $\mathbf{B}' = \mathbf{A}$. Lastly if a matrix consists of only one element it is known as a '*scalar*' and is normally written in small Greek letters. Thus the ordinary algebra we use is 'scalar algebra'.

We said that matrix algebra was developed to deal with the algebra of simultaneous linear equations. We can write any such set of equations in the form

$$A_{11}x_1 + \ldots + A_{1K}x_K = b_1$$
$$\vdots \qquad\qquad \vdots \qquad \vdots \tag{1}$$
$$A_{N1}x_1 + \ldots + A_{NK}x_K = b_N$$

in scalar algebra. For example

$$2x_1 + 3x_2 = 7$$
$$4x_1 + 2x_2 = 10$$

where

$$\mathbf{A} = \begin{bmatrix} 2 & 3 \\ 4 & 2 \end{bmatrix}, \quad \mathbf{x} = \begin{bmatrix} x_1 \\ x_2 \end{bmatrix}, \quad \mathbf{b} = \begin{bmatrix} 7 \\ 10 \end{bmatrix}$$

The object of the exercise is to solve these equations for x. In matrix algebra we can write the equations as

$$\mathbf{Ax} = \mathbf{b} \tag{2}$$

and in the same way as we would solve

$$\alpha\gamma = \beta$$

for γ as $\gamma = \alpha^{-1}\beta$ we can solve (2) as

$$\mathbf{x} = \mathbf{A}^{-1}\mathbf{b} \tag{3}$$

This gives us two further matrix operations, multiplication, $\mathbf{Ax}$, and 'division' or *inversion* as it is called, the calculation of $\mathbf{A}^{-1}$. Firstly perhaps we should mention the simpler operations of addition and subtraction; if $\mathbf{C} = \mathbf{A} + \mathbf{B} - \mathbf{D}$, $C_{ij} = A_{ij} + B_{ij} - D_{ij}$ for all i and j, where all four matrices must have the same dimensions. If we look at equation (2) we can see what multiplication must involve. A is an $N \times K$ matrix, x is a vector of length K and b is a vector of length N. Hence Ax must be a vector of length N if Ax = b. We know that

$$b_1 = \sum_k A_{1k} x_k, \quad b_2 = \sum_k A_{2k} x_k$$

and so on from (2). In general if

$$AB = C \tag{4}$$

$$C_{ij} = \sum_k A_{ik} B_{kj} \tag{5}$$

Thus C must have the same number of rows as A and the same number of columns as B, and A must have the same number of columns as B has rows. We can check that this is true of (2): b has the same number of rows as A, namely N, and b has the same number of columns as x, namely 1, and A has the same number of columns as x has rows; they are thus what is termed 'conformable'. It is worth noting that AB will not normally equal BA, in fact they may not be conformable in the reverse order.

The last step of inversion is much more difficult to calculate; however, it is easy to define. In the same way that

$$\alpha^{-1} \alpha = 1 = \alpha \alpha^{-1}$$

$$A^{-1} A = I = AA^{-1}$$

where I is an '*identity*' matrix, a matrix which is square and has one on the main diagonal and zero elsewhere

$$I = \begin{bmatrix} 1 & 0 & 0 \ldots 0 \\ 0 & 1 & 0 \ldots 0 \\ 0 & 0 & 1 \cdot \cdot \vdots \\ \vdots & \vdots \cdot \cdot \cdot 0 \\ 0 & 0 \ldots 0 & 1 \end{bmatrix}$$

Since $A^{-1} A = AA^{-1}$ it follows that A and A^{-1} must be square and have the same dimensions. The only further condition is that all the rows (or columns) in A must be linearly independent, where the rows of A are said to be linearly independent if the only solution to Ab = 0 is b = 0.

There are many ways of finding inverses but this in general is beyond the scope of this book. However, let us take the simple two-equation case where, for example

$$2x_1 + 3x_2 = 7 \tag{6}$$

$$4x_1 + 2x_2 = 10 \tag{7}$$

Solving, multiply (6) by ½

$$x_1 + 3/2x_2 = 7/2 \tag{8}$$

Add −4 times (8) to (7)

$$-4x_2 = -4 \tag{9}$$

Multiply (9) by −¼

$$x_2 = 1 \tag{10}$$

Add −3/2 times (10) to (8)

$$x_1 = 2 \tag{11}$$

This method of solution is known as Gauss—Jordan elimination. It can be generalized for a matrix of any size; in effect one solves each x_i in turn in terms of the remaining xs and substitutes that value back into the other equations.

We can find A^{-1} rather than $A^{-1}b$ by performing exactly the same operations on an ($N \times N$) identity matrix as we did on equations (6) and (7). In this case N is 2, therefore

$$I = \begin{bmatrix} 1 & 0 \\ 0 & 1 \end{bmatrix}$$

Multiply row 1 by ½

$$\begin{bmatrix} ½ & 0 \\ 0 & 1 \end{bmatrix}$$

Add −4 times row 1 to row 2

$$\begin{bmatrix} ½ & 0 \\ -2 & 1 \end{bmatrix}$$

Multiply row 2 by −¼

$$\begin{bmatrix} ½ & 0 \\ ½ & -¼ \end{bmatrix}$$

Add −3/2 times row 2 to row 1

$$\begin{bmatrix} -¼ & ^3/_8 \\ ½ & -¼ \end{bmatrix}$$

To check that this is the inverse

$$\begin{bmatrix} -¼ & ^3/_8 \\ ½ & -¼ \end{bmatrix} \begin{bmatrix} 2 & 3 \\ 4 & 2 \end{bmatrix} = \begin{bmatrix} 1 & 0 \\ 0 & 1 \end{bmatrix}$$

Finally to solve

$$x = A^{-1} b$$

$$x = \begin{bmatrix} -\tfrac{1}{4} & \tfrac{3}{8} \\ \tfrac{1}{2} & -\tfrac{1}{4} \end{bmatrix} \begin{bmatrix} 7 \\ 10 \end{bmatrix}$$

Recalling (5) we can write that

$$x = \begin{bmatrix} A_{11}^{-1} b_1 + A_{21}^{-1} b_2 \\ A_{12}^{-1} b_1 + A_{22}^{-1} b_2 \end{bmatrix}$$

$$= \begin{bmatrix} (\tfrac{1}{4})(7) + (\tfrac{3}{8})(10) \\ (\tfrac{1}{2})(7) + (-\tfrac{1}{4})(10) \end{bmatrix} = \begin{bmatrix} 2 \\ 1 \end{bmatrix}$$

which is exactly what we obtained in (10) and (11).

With this knowledge let us return to the Normal Equations. They are clearly a set of simultaneous equations, and they run as follows:

$$
\begin{aligned}
N\hat{b}_0 \;\;\;\;\;\; + \hat{b}_1 \Sigma X_{i1} \;\;\;\;\; + \ldots + \hat{b}_K \Sigma X_{iK} &= \Sigma y_i \\
\hat{b}_0 \Sigma X_{i1} + \hat{b}_1 \Sigma X_{i1}^2 \;\;\;\;\; + \ldots + \hat{b}_K \Sigma X_{i1} X_{iK} &= \Sigma X_{i1} y_i \\
\hat{b}_0 \Sigma X_{i2} + \hat{b}_1 \Sigma X_{i2} X_{i1} + \ldots + \hat{b}_K \Sigma X_{i2} X_{iK} &= \Sigma X_{i2} y_i \\
\vdots \qquad\qquad \vdots \qquad\qquad\qquad \vdots \qquad\qquad\qquad \vdots \\
\hat{b}_0 \Sigma X_{iK} + \hat{b}_1 \Sigma X_{iK} X_{i1} + \ldots + \hat{b}_K \Sigma X_{iK}^2 &= \Sigma X_{iK} y_i
\end{aligned}
\tag{12}
$$

Our first step is to sort them out into matrices and vectors so that we can express them in the form $Ax = b$. With the exception of the very first element these equations are concerned with one matrix and two vectors. The matrix is X and the vectors $\hat{b}$ and y. $\hat{b}$ is the vector of $K + 1$ least-squares coefficients, y is the vector of N observations on the endogenous variable and X is the matrix of N observations (rows) on the K exogenous variables (columns). We can include the first element in the same framework if we add a new first column to X which consists entirely of ones.

$$X = \begin{bmatrix} 1 & X_{11} & X_{12} & \ldots X_{1K} \\ 1 & X_{21} & X_{22} & \ldots X_{2K} \\ \vdots & \vdots & \vdots & \vdots \\ 1 & X_{N1} & X_{N2} & \ldots X_{NK} \end{bmatrix}$$

we can now rewrite the equation as

$$
\begin{aligned}
\hat{b}_0 \Sigma X_{i0}^2 \;\;\;\;\; + \hat{b}_1 \Sigma X_{i0} X_{i1} + \ldots + \hat{b}_K \Sigma X_{i0} X_{iK} &= \Sigma X_{i0} y_i \\
\hat{b}_0 \Sigma X_{i1} X_{i0} + \hat{b}_1 \Sigma X_{i1}^2 \;\;\;\;\; + \ldots + \hat{b}_K \Sigma X_{i1} X_{iK} &= \Sigma X_{i1} y_i \\
\vdots \qquad\qquad \vdots \qquad\qquad\qquad \vdots \qquad\qquad \vdots \\
\hat{b}_0 \Sigma X_{iK} X_{i0} + \hat{b}_1 \Sigma X_{iK} X_{i1} + \ldots + \hat{b}_K \Sigma X_{iK}^2 &= \Sigma X_{iK} y_i
\end{aligned}
\tag{13}
$$

where X_0 is the first column of $\mathbf{X}$. We can see immediately that $\Sigma X_{i0}^2 = \Sigma 1^2 = N$, $\Sigma X_{i0} X_{i1} = \Sigma 1 X_{i1} = \Sigma X_{i1}$, and so on, so that the values of each term in (12) and (13) are identical.

It is quite easy to express the right-hand side of (13) in matrix notation as $\mathbf{X'y}$. The left-hand side is more difficult, but is clearly some multiple of $\mathbf{X}$ and $\hat{\mathbf{b}}$. It is in fact $\mathbf{X'X\hat{b}}$. Let us check this by looking at say the first element of the first equation, $\hat{b}_0 \Sigma X_{i0}^2$. This is equal to $\hat{b}_0$ multiplied into the sum of each element in the first row of $\mathbf{X'}$ multiplied by the respective element in the first column of $\mathbf{X}$. The other terms on the left-hand side of the first equation of (13) are the elements in $\hat{\mathbf{b}}$ multiplied by the respective elements in the first row of $\mathbf{X'X}$. Hence the whole left-hand side is $\mathbf{X'X\hat{b}}$, and equating this with the right-hand side gives

$$\mathbf{X'X\hat{b}} = \mathbf{X'y} \tag{14}$$

Finally solving for $\hat{\mathbf{b}}$ we obtain

$$\hat{\mathbf{b}} = (\mathbf{X'X})^{-1} \mathbf{X'y} \tag{15}$$

It is abundantly clear that (14) and (15) are very much simpler ways of referring to the equations than scalar algebra.

This completes the matrix algebra we require. There are many books that the interested reader can consult if he wishes to take this further (see for example G. Mills, G. Hadley). We can see how easy it is to manipulate matrix algebra, it is only the computation of inverses which is difficult, and we would use a standard routine on a computer to calculate any but the very simplest examples.

Suggested Reading
See Chapter 7.

Appendix 3 Statistical Tables

Table A3.1
Random Numbers

6450	1787	8896	9255	7188	8236	1333	2800
0108	3468	3210	0964	2634	1184	7217	5350
4405	9374	3212	6411	2380	6629	7742	8163
2774	3275	6363	3714	1762	9281	0757	3580
7017	0988	2657	8806	2008	8624	6563	0879
5144	6622	3260	0463	5671	8745	6143	6901
8411	0006	2336	0550	9271	7384	0612	9893
3565	1345	8107	5461	9138	0166	7850	2553
6280	9048	9076	4758	6451	6044	2347	0270
1797	6772	1476	3011	7474	9940	4243	0555
8765	5663	3912	0837	5002	8583	3561	9689
2282	7793	5191	1143	6995	5552	5655	2562
2134	6806	1370	6560	8424	1510	3846	1714
2235	1750	7997	4090	2307	5655	5266	1567
7266	0124	3543	2946	5947	4344	5889	3877
2514	6112	0052	1581	8381	2949	0785	0365
2921	0028	6973	9944	5008	6887	5545	8561
7312	4954	4201	2907	5438	7864	4931	6841
3838	6580	0326	0911	0530	1313	4711	4672
2621	2243	2058	3808	4628	1036	2706	4047
1592	5169	0880	3893	9007	7043	5019	8130
1529	9911	2882	4153	2507	2590	3483	3086
6297	4991	7802	7702	5379	6425	0303	9137
4289	8738	3263	4420	2321	6220	5886	6786
7066	4331	0221	0797	0717	9217	5889	4266
1198	0034	5597	8973	0081	6065	3460	2181
9305	0641	8121	0969	8696	3863	2670	1340
6290	6115	3368	4143	3450	4395	9163	9800
6787	3424	4007	3816	8881	3575	3988	5107
9798	7586	1231	9116	1414	8085	6641	7734
6533	7907	3402	8020	4805	5214	3309	1721
7449	5420	7887	7595	2773	1901	6305	8523

Table A3.2

Normal Probability Tables. Table shows probability of a value of Z being greater than Z_c where Z is distributed $N(0,1)$

Z_c	0.00	0.01	0.02	0.03	0.04	0.05	0.06	0.07	0.08	0.09
0.0	0.5000	0.4960	0.4920	0.4880	0.4840	0.4801	0.4761	0.4721	0.4681	0.4641
0.1	0.4602	0.4562	0.4522	0.4483	0.4443	0.4404	0.4364	0.4325	0.4286	0.4247
0.2	0.4207	0.4168	0.4129	0.4090	0.4052	0.4013	0.3974	0.3936	0.3897	0.3859
0.3	0.3821	0.3783	0.3745	0.3707	0.3669	0.3632	0.3594	0.3557	0.3520	0.3483
0.4	0.3446	0.3409	0.3372	0.3336	0.3300	0.3264	0.3228	0.3192	0.3156	0.3121
0.5	0.3085	0.3050	0.3015	0.2981	0.2946	0.2912	0.2877	0.2843	0.2810	0.2776
0.6	0.2743	0.2709	0.2676	0.2643	0.2611	0.2578	0.2546	0.2514	0.2483	0.2451
0.7	0.2420	0.2389	0.2358	0.2327	0.2296	0.2266	0.2236	0.2206	0.2177	0.2148
0.8	0.2119	0.2090	0.2061	0.2033	0.2005	0.1977	0.1949	0.1922	0.1894	0.1867
0.9	0.1841	0.1814	0.1788	0.1762	0.1736	0.1711	0.1685	0.1660	0.1635	0.1611
1.0	0.1587	0.1562	0.1539	0.1515	0.1492	0.1469	0.1446	0.1423	0.1401	0.1379
1.1	0.1357	0.1335	0.1314	0.1292	0.1271	0.1251	0.1230	0.1210	0.1190	0.1170
1.2	0.1151	0.1131	0.1112	0.1093	0.1075	0.1056	0.1038	0.1020	0.1003	0.0985
1.3	0.0968	0.0951	0.0934	0.0918	0.0901	0.0885	0.0869	0.0853	0.0838	0.0823
1.4	0.0808	0.0793	0.0778	0.0764	0.0749	0.0735	0.0721	0.0708	0.0694	0.0681
1.5	0.0668	0.0655	0.0643	0.0630	0.0618	0.0606	0.0594	0.0582	0.0571	0.0559
1.6	0.0548	0.0537	0.0526	0.0516	0.0505	0.0495	0.0485	0.0475	0.0465	0.0455
1.7	0.0446	0.0436	0.0427	0.0418	0.0409	0.0401	0.0392	0.0384	0.0375	0.0367
1.8	0.0359	0.0351	0.0344	0.0336	0.0329	0.0322	0.0314	0.0307	0.0301	0.0294
1.9	0.0287	0.0281	0.0274	0.0268	0.0262	0.0256	0.0250	0.0244	0.0239	0.0233
2.0	0.0228	0.0222	0.0217	0.0212	0.0207	0.0202	0.0197	0.0192	0.0188	0.0183
2.1	0.0179	0.0174	0.0170	0.0166	0.0162	0.0158	0.0154	0.0150	0.0146	0.0143
2.2	0.0139	0.0136	0.0132	0.0129	0.0125	0.0122	0.0119	0.0116	0.0113	0.0110
2.3	0.0107	0.0104	0.0102	0.0099	0.0096	0.0094	0.0091	0.0089	0.0087	0.0084
2.4	0.0082	0.0080	0.0078	0.0075	0.0073	0.0071	0.0069	0.0068	0.0066	0.0064
2.5	0.0062	0.0060	0.0059	0.0057	0.0055	0.0054	0.0052	0.0051	0.0049	0.0048
2.6	0.0047	0.0045	0.0044	0.0043	0.0041	0.0040	0.0039	0.0038	0.0037	0.0036
2.7	0.0035	0.0034	0.0033	0.0032	0.0031	0.0030	0.0029	0.0028	0.0027	0.0026
2.8	0.0026	0.0025	0.0024	0.0023	0.0023	0.0022	0.0021	0.0021	0.0020	0.0019
2.9	0.0019	0.0018	0.0018	0.0017	0.0016	0.0016	0.0015	0.0015	0.0014	0.0014
3.0	0.0013	0.0013	0.0013	0.0012	0.0012	0.0011	0.0011	0.0011	0.0010	0.0010
3.1	0.0010	0.0009	0.0009	0.0009	0.0008	0.0008	0.0008	0.0008	0.0007	0.0007

Table A3.3
Student's t distribution. Table shows probability of a value t being greater than t_c where t is distributed with DF degrees of freedom

DF	0.10	0.05	0.025	0.01	0.005
5	1.48	2.01	2.56	3.35	4.03
6	1.44	1.94	2.44	3.14	3.72
7	1.42	1.89	2.36	3.00	3.50
8	1.40	1.86	2.31	2.90	3.36
9	1.38	1.83	2.26	2.82	3.25
10	1.37	1.81	2.23	2.76	3.17
11	1.36	1.80	2.20	2.72	3.11
12	1.36	1.78	2.18	2.68	3.05
13	1.35	1.77	2.16	2.65	3.01
14	1.35	1.76	2.14	2.62	2.98
15	1.34	1.75	2.13	2.60	2.95
16	1.34	1.75	2.12	2.58	2.92
17	1.33	1.74	2.11	2.57	2.90
18	1.33	1.73	2.10	2.55	2.88
19	1.33	1.73	2.09	2.54	2.86
20	1.33	1.72	2.09	2.53	2.85
21	1.32	1.72	2.08	2.52	2.83
22	1.32	1.72	2.07	2.51	2.82
23	1.32	1.71	2.07	2.50	2.81
24	1.32	1.71	2.06	2.49	2.80
25	1.32	1.71	2.06	2.49	2.79
30	1.31	1.70	2.04	2.46	2.75
40	1.30	1.68	2.02	2.42	2.71
50	1.30	1.68	2.01	2.40	2.68
∞	1.28	1.65	1.96	2.33	2.58

Table A3.4

Table of the F variance ratio. Table gives values of F_c where probabilities of a value F exceeding F_c are 5 percent and 1 percent with degrees of freedom DF_1 and DF_2

DF_2	DF_1 1		2		3		4		5		6		7		8		9		10	
	5%	1%	5%	1%	5%	1%	5%	1%	5%	1%	5%	1%	5%	1%	5%	1%	5%	1%	5%	1%
5	6.61	16.26	5.79	13.27	5.41	12.06	5.19	11.39	5.05	10.97	4.95	10.67	4.88	10.45	4.82	10.27	4.78	10.15	4.74	10.05
6	5.99	13.74	5.14	10.92	4.76	9.73	4.53	9.16	4.39	8.75	4.28	8.47	4.21	8.26	4.15	8.10	4.10	7.98	4.06	7.87
7	5.59	12.25	4.74	9.55	4.35	8.45	4.12	7.85	3.97	7.46	3.87	7.19	3.79	7.00	3.73	6.84	3.68	6.71	3.63	6.62
8	5.32	11.26	4.46	8.65	4.07	7.59	3.84	7.01	3.69	6.63	3.58	6.37	3.50	6.19	3.44	6.03	3.39	5.91	3.34	5.82
9	5.12	10.56	4.26	8.02	3.86	6.99	3.63	6.42	3.48	6.06	3.37	5.80	3.29	5.62	3.23	5.47	3.18	5.35	3.13	5.26
10	4.96	10.04	4.10	7.56	3.71	6.55	3.48	5.99	3.33	5.64	3.22	5.39	3.14	5.21	3.07	5.06	3.02	4.95	2.97	4.85
11	4.84	9.65	3.98	7.20	3.59	6.22	3.36	5.67	3.20	5.32	3.09	5.07	3.01	4.88	2.95	4.74	2.90	4.63	2.86	4.54
12	4.75	9.33	3.88	6.93	3.49	5.95	3.26	5.41	3.11	5.06	3.00	4.82	2.92	4.65	2.85	4.50	2.80	4.39	2.76	4.30
13	4.67	9.07	3.80	6.70	3.41	5.74	3.18	5.20	3.02	4.86	2.92	4.62	2.84	4.44	2.77	4.30	2.72	4.19	2.67	4.10
14	4.60	8.86	3.74	6.51	3.34	5.56	3.11	5.03	2.96	4.69	2.85	4.46	2.77	4.28	2.70	4.14	2.65	4.03	2.60	3.94
15	4.54	8.68	3.68	6.36	3.29	5.42	3.06	4.89	2.90	4.56	2.79	4.32	2.70	4.14	2.64	4.00	2.59	3.89	2.55	3.80
16	4.49	8.53	3.63	6.23	3.24	5.29	3.01	4.77	2.85	4.44	2.74	4.20	2.66	4.03	2.59	3.89	2.54	3.78	2.49	3.69
18	4.41	8.28	3.55	6.01	3.16	5.09	2.93	4.53	2.77	4.25	2.66	4.01	2.58	3.85	2.51	3.71	2.46	3.60	2.41	3.51
20	4.35	8.10	3.49	5.85	3.10	4.94	2.87	4.43	2.71	4.10	2.60	3.87	2.52	3.71	2.45	3.56	2.40	3.45	2.35	3.37
22	4.30	7.94	3.44	5.72	3.05	4.82	2.82	4.31	2.66	3.99	2.55	3.76	2.47	3.59	2.40	3.45	2.35	3.35	2.30	3.26
25	4.24	7.77	3.38	5.57	2.99	4.68	2.76	4.18	2.60	3.86	2.49	3.63	2.41	3.46	2.34	3.32	2.28	3.21	2.24	3.13
30	4.17	7.56	3.32	5.39	2.92	4.51	2.69	4.02	2.53	3.70	2.42	3.47	2.34	3.30	2.27	3.17	2.21	3.06	2.16	2.98
40	4.08	7.31	3.23	5.18	2.84	4.31	2.61	3.83	2.45	3.51	2.34	3.29	2.25	3.12	2.18	2.99	2.12	2.88	2.07	2.80
60	4.00	7.08	3.15	4.98	2.76	4.13	2.53	3.65	2.37	3.34	2.25	3.12	2.17	2.95	2.10	2.82	2.04	2.72	1.99	2.63
120	3.92	6.85	3.07	4.79	2.68	3.95	2.45	3.48	2.29	3.17	2.18	2.96	2.09	2.79	2.02	2.66	1.96	2.56	1.91	2.47

Derived from Table 18 of Biometrika Tables for Statisticians, Volume 1 with permission of the Trustees.

Table A3.5

Table of Durbin–Watson d. The 5% and 1% significance points are shown for the distributions of d_L and d_U the lower and upper bounds of d for given numbers of observations N and numbers of exogeneous variables K (excluding the constant) in the regression

N	$K = 1$		$K = 2$		$K = 3$		$K = 4$		$K = 5$	
	d_L	d_U	d_L	d_U	d_L	d_U	d_L	d_U	d_L	d_U
					At the 5% level of significance					
15	1.08	1.36	0.95	1.54	0.82	1.75	0.69	1.97	0.56	2.21
20	1.20	1.41	1.10	1.54	1.00	1.68	0.90	1.83	0.79	1.99
25	1.29	1.45	1.21	1.55	1.12	1.66	1.04	1.77	0.95	1.89
30	1.35	1.49	1.28	1.57	1.21	1.65	1.14	1.74	1.07	1.83
35	1.40	1.52	1.34	1.58	1.28	1.65	1.22	1.73	1.16	1.80
40	1.44	1.54	1.39	1.60	1.34	1.66	1.29	1.72	1.23	1.79
50	1.50	1.59	1.46	1.63	1.42	1.67	1.38	1.72	1.34	1.77
100	1.65	1.69	1.63	1.72	1.61	1.74	1.59	1.76	1.57	1.78
					At the 1% level of significance					
15	0.81	1.07	0.70	1.25	0.59	1.46	0.49	1.70	0.39	1.96
20	0.95	1.15	0.86	1.27	0.77	1.41	0.68	1.57	0.60	1.74
25	1.05	1.21	0.98	1.30	0.90	1.41	0.83	1.52	0.75	1.65
30	1.13	1.26	1.07	1.34	1.01	1.42	0.94	1.51	0.88	1.61
35	1.19	1.31	1.14	1.37	1.08	1.44	1.03	1.51	0.97	1.59
40	1.25	1.34	1.20	1.40	1.15	1.46	1.10	1.52	1.05	1.58
50	1.32	1.40	1.28	1.45	1.24	1.49	1.20	1.54	1.16	1.59
100	1.52	1.56	1.50	1.58	1.48	1.60	1.46	1.63	1.44	1.65

Derived from Biometrilca, Volume 38, pages 173 and 175, with permission of the Trustees.

Index